CW00468739

UELI SEILER-HUGOVA (b. 1942) attended a Steiner-Waldorf school, trained as a primary school teacher and studied at Berne University and the Zurich Institute for Applied Psychology. He was head teacher of a state school from 1972 to 2006, and is currently director of the Independent Seminar for Institutional Education at Schlössli Ins, Switzerland. He is also a guest professor for education at the University of Latvia, Riga, a teacher at the Technical University for Artistic Social Therapy in Prague, and collaborates with the Research Project on Integral Education at Regensburg University.

Ueli Seiler-Hugova gives lectures and courses on Goethe's theory of colour, on the theory of the senses according to Rudolf Steiner and Hugo Kükelhaus, and on integral star studies (astronomy, astrology and astrosophy).

Colour

Seeing, Experiencing, Understanding

Ueli Seiler-Hugova

TEMPLE LODGE

Dedicated with gratitude to Kamila, Julian, Manuel and Alma

Temple Lodge Publishing
Hillside House, The Square
Forest Row, RH18 5ES

www.templelodge.com

Published by Temple Lodge 2011

Originally published in German under the title *Farben, sehen, erleben, verstehen* by AT Verlag, Aarau, Switzerland in 2002. This edition is based on the revised second edition, 2007

A catalogue record for this book is available from the British Library

ISBN 978 1 906999 23 0

Cover by Andrew Morgan Design
Typeset by DP Photosetting, Neath, West Glamorgan
Printed and bound by Gutenberg Press Ltd., Malta

The paper used for this book is FSC-certified and totally chlorine-free. FSC (the Forest Stewardship Council) is an international network to promote responsible management of the world's forests.

CONTENTS

FOREWORD

'I and the colours are one'

Paul Klee

Archetypal phenomenon:
ideal, real, symbolic, identical.

Ideal as the ultimate we can know;
real as what we know;
symbolic, because it includes all instances;
identical with all instances.

J. W. von Goethe

This book sets out to describe colours. It describes how they come into being – both out of darkness and out of light. The path out of darkness into light and out of light into darkness described by Johann Wolfgang von Goethe is one that we can all experience. We have to perceive the colours independently, for only then can we make them a part of ourselves.

By perceiving colours in this way we are led to a holistic image of them – a colour circle in which impressionistic perception meets up with expressionistic interpretation of the phenomena. Perceptions ask to be interpreted, named and explained. Even as little children we want to give a name to every single thing we see.

This book on colour aims for a holistic perception and an all-embracing interpretation. The chief authorities cited are Johann Wolfgang von Goethe, Vincent van Gogh, Rudolf Steiner and the colour researcher Harald Küppers.

The approach is not scientific in the general sense, so there are no detailed source data. However, an ample bibliography points to works that can help deepen one's research. The aim is simply to point out a path, a colour path that has been trodden by the author countless times in the company of course participants. Being meditative in character, it describes not only outer but also inner observations. The intention is to give an introduction to the subject of colour which is simple and easily understood and which in the end can lead to a survey of colours, a perception of colours and their meaning.

We are concerned neither with Newton's wave theory nor with any more modern views such as that of quantum physics, for the book's intention is to be complementary, to provide a supplement to present-day theories about colour. Goethe's way of looking at colours provides a convenient basis for an introduction to the subject. It is useful above all in teaching non-scientists, children, youngsters, schoolteachers, autodidacts, grandmothers and grandfathers.

At a youth conference at the Goetheanum centre in Switzerland in 1963 it was Heinrich O. Proskauer who first introduced me to the world of prismatic colours and coloured shadows. Since then I have never ceased in my researches while developing my own theory of colours in courses for schoolchildren and students – courses in which I, too, have gained a great variety of new insights along the way.

My thanks also go to retired schoolteacher Fritz Berger for interesting encounters with colour. In the sixth class of the Berne Rudolf Steiner School he was the first to stimulate my fascination with the subject. The book has gained its reference to orthodox colour research, for example the phenomenon of additive colours, through Ueli Aeschlimann who guided a group of schoolteachers in their study of colours. And in working with Elisabeth Aeschbach I discovered many practical insights into the perception of colour. It was with her that I designed my first colour circles using natural materials. Many other personal encounters and discoveries in colour literature have also contributed to the creation of this book.

To my family I owe the possibility vouchsafed me in the summer of 1998 to inhabit a deserted house at Velke Mezirici (Czech Republic). There I found the necessary space in which to write down the main draft of the book which had for years been maturing both in my head and in countless written notes. A generous donation from my godmother, Hilde Madliger, enabled me to drive the project forward in the spring of 1999, and that summer the draft was edited by Vanda Messerli. Pavel Selesi from Prague then put in train the initial printed draft. Here it was my Czech wife Kamila who helped as an interpreter in many discussions with the graphic designer. Most of the photos were created by René Bürgy. And finally Otto Schärli introduced me to the original (Swiss) publisher, AT Verlag. It was this publisher who made it possible for the book to be produced. My most heartfelt thanks go to all the above for their assistance.

This second, considerably expanded edition has an extra introductory chapter (Experiencing the world of colours) in which experiments with rainbows, coloured shadows, after-image colours (complementary colours) and soap bubbles are described. Without having to refer in any way to the subsequent chapters, readers can study, envisage, experience and hopefully also understand the phenomena with the help of the examples given.

The phenomena discussed in this initial chapter were pointed out to me by many friends. Special mention is due to Ueli Bühler, director of the care home at Brüttelen, Switzerland, who introduced me to techniques for observing colours in soap bubbles. And I thank Nicolas Kyramarios in Berne above all for the very first photographs of the total, 360-degree, rainbow, a phenomenon which could of course be seen subjectively by an observer but is now made visible objectively in these photos.

Warm thanks go to the publisher for enabling this considerably expanded second edition to appear in print.

Ueli Seiler-Hugova

EXPERIENCING THE WORLD OF COLOURS

The following experiments with rainbows, coloured shadows, complementary after-image colours and soap bubbles demonstrate simple ways of having fun with colour phenomena.

Rainbows

Observing rainbows in the sky

We have all seen a rainbow in the sky at one time or another. As soon as the sun comes out again after rain, stand with your back to the sun and look in the direction of the departing cloud.

Here is a beautiful rainbow with the sky showing dark above it and light beneath. Sometimes additional rainbows also appear on the underside. Then through addition the colour magenta will arise between the violet and the vermilion of the lower rainbow.

This picture shows a main rainbow with a secondary, fainter rainbow outside it in which the colour sequence is reversed.

A rainbow generated with the help of the rainbow instrument. The astrolabe, for use in observing sun, moon and stars, is seen in the distance.

A complete 360-degree rainbow is another achievement of the rainbow instrument, here with the observer's shadow in the centre, photographed using a fisheye lens.

Rainbow instruments

Rainbows can be produced experimentally either with the help of the sun or at night with the help of a projector.

A static rotating rainbow instrument

Beside a pond in the Rosenhof Park of the educational establishment Schlössli Ins in Switzerland there is a device by means of which one can generate a wonderful rainbow with the help of flowing water and sunshine (see also pages 51 & 52).

Water flows upwards in a fixed vertical pipe and outwards into a horizontal one from which it then descends vertically through nozzles. To generate a rainbow the rays of the sun must strike the curtain of water at right angles. The horizontal water pipe can be rotated as necessary.

This rainbow screen shows texts referring to Goethe's theory of colours: 'Every individual has his/her own rainbow.' – 'In the coloured reflections we see life.' The screen was designed by the author and manufactured by the firm of Hofer at Müntschemier, Switzerland.

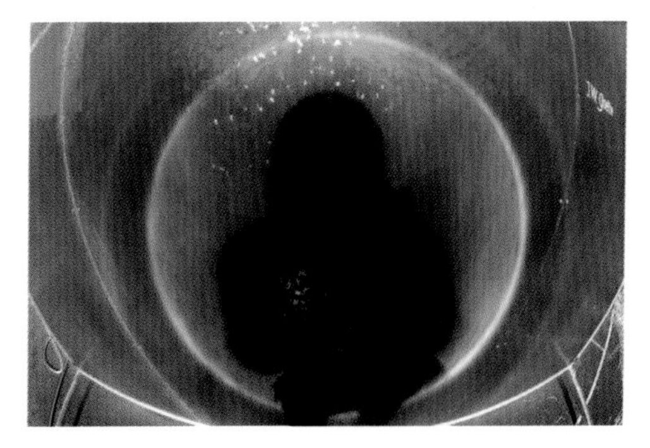

Photo of a complete 360-degree rainbow on the rainbow screen with the observer's shadow in the centre, again photographed with a fisheye lens.

Rotating rainbow screen

A horizontal water pipe sending a curtain of water down from nozzles is fixed along the top edge of a movable screen (3 × 3m). The screen is then turned to stand at right angles to the rays of sunlight. When the sun is low in the sky a shadow of the observer's head will appear in the centre of the 360-degree rainbow circle if he stands close to the curtain of water. When the sun is higher in the sky a complete rainbow can only be seen when the observer stands on a raised platform or chair.

Photographing rainbows

Only parts of a rainbow can be photographed with a normal camera. But with a fisheye lens you can take a picture of a complete 360-degree rainbow circle. This depends on the 42-degree angular separation as seen from the position of the shadow of the observer's head. The angular separation has to be at least doubled in order to obtain a sufficiently large angle in the lens.

The distance from the rainbow screen also has a part to play. In the case of a relatively small screen (e.g. 3 × 3m) one must stand not more than 30cm from the screen in order to see the full rainbow.

The fact that every individual sees his/her own rainbow is an expression of *subjective objectivity*: although exact optical laws apply, the phenomenon can only be seen by the individual.

Rainbows at night

If the rainbow screen (see p.7) is photographed at night or in the dark when lit by a projector you also see a rainbow. During the 'Museum at Night' event which took place from 6pm to midnight in front of the Federal Government House in Berne on 23 March 2007, over a thousand people were able to look at a rainbow on the rainbow screen. In some instances they had to wait for up to twenty minutes to see the 'miracle'.

For the 'Museum at Night' event the rainbow screen was enclosed between two walls and an overhang each measuring 3 × 3m, thus forming the inside of a cube. Posters describing optical phenomena were displayed at the sides. The rainbow appeared in the form of a circle surrounding a shadow of the observer's head when he/she stood sufficiently close to the curtain of water.

The rainbow instrument in front of Government House with the 5000-Watt Freshnel projector.

A rainbow in the sky photographed with a fisheye lens.

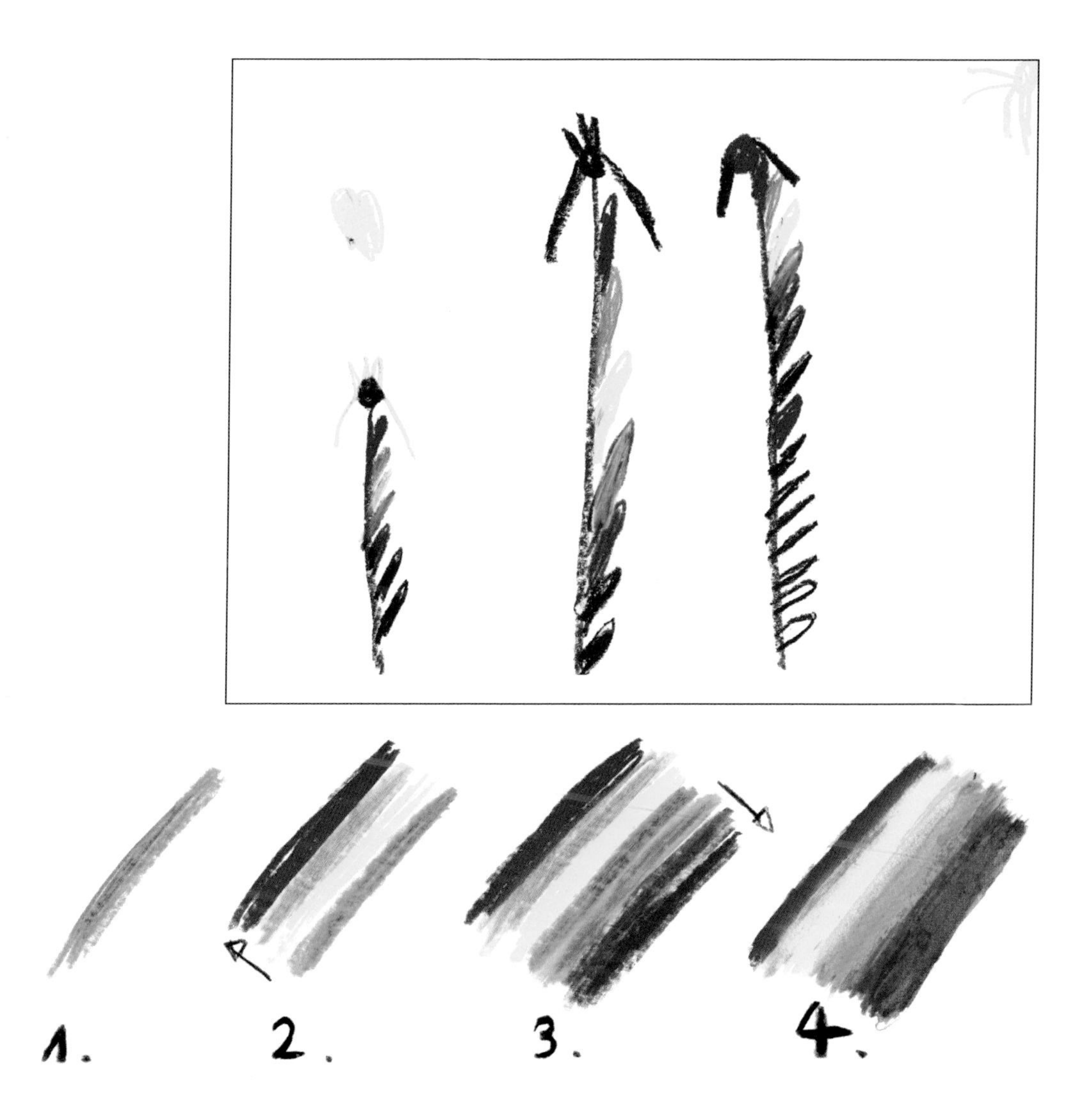

Painting and drawing rainbows

Painting rainbows is one of the most delightful activities for children. They are overjoyed once they have learnt the correct sequence of the colours by heart.

In coming to grips with the sequence of the seven rainbow colours it is a help for children, as well as adults, to begin with the middle colour, green.

Having begun with green (1), add yellow, orange and vermilion going outwards (2). Then draw pale blue, indigo (deep blue) and finally violet going inwards (3). Caran d'Ache, Neocolor II water soluble crayons can be painted over with water to obtain a watercolour effect (4).

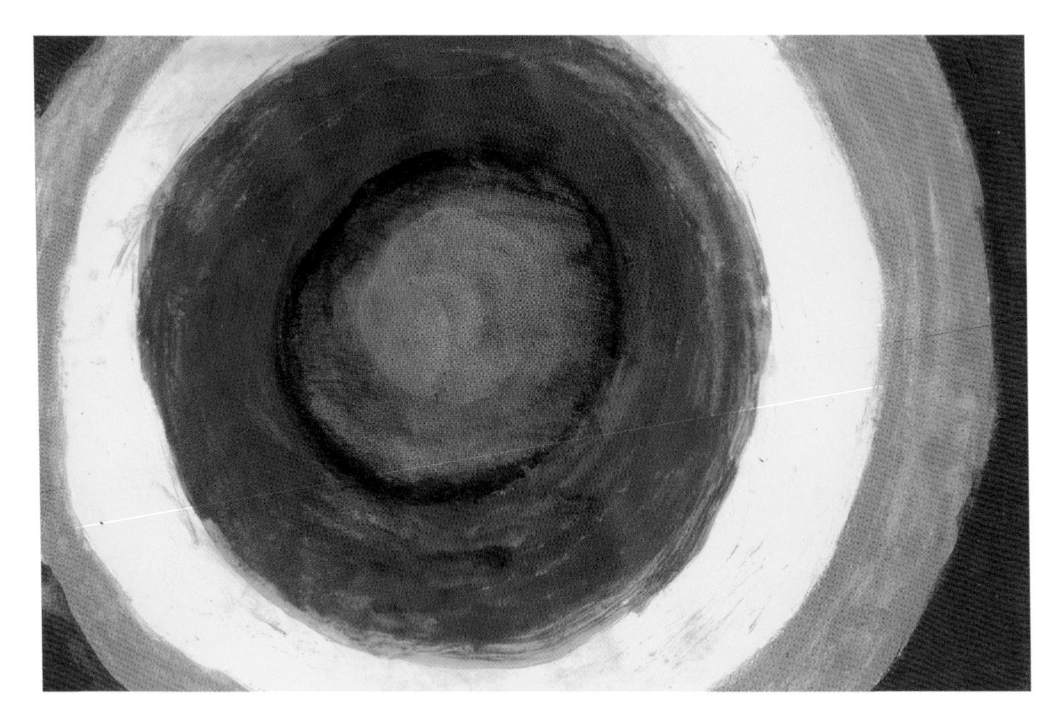

A 'Brocken spectre', i.e. the observer's shadow surrounded by a rainbow, is seen looking down from a mountainside on to mist spread out below.

Coloured shadows

Experimenting with coloured shadows

Otto von Guericke described the phenomenon of coloured shadows as early on as 1672. Goethe considered they belonged to the physiological colours; in other words he assumed that coloured shadows arose in the eye of the beholder in the same way as the after-image colours. Rudolf Steiner, however, saw them as arising in the external world.

Four projectors are needed in order to generate coloured shadows, three to project colours and one for white. The colours needed are the ones that are produced by the retina in the eye, or those on which coloured television pictures are based. These are: vermilion, violet-blue and green.

The best results are obtained in our experiments if the brightness of the light projected can be varied.

Green snow

What we see in this experiment is described as follows by Goethe in his Theory of Colours (from *Goethe, The Collected Works, Vol.12, Scientific Studies*, Princeton University Press, translated from German by Douglas Miller):

'Once, on a winter's journey in the Harz Mountains, I made my descent from the Brocken as evening fell. The broad slope above and below me was snow-covered, the meadow lay beneath a blanket of snow, every isolated tree and jutting crag, every wooded grove and rocky prominence was rimed with frost, and the sun was just setting beyond the Oder ponds.

'Because of the snow's yellowish cast, pale violet shadows had accompanied us all day, but now, as an intensified yellow reflected from the areas in the light, we were obliged to describe the shadows as deep blue.

'At last the sun began to disappear and its rays, subdued by the strong haze, spread the most beautiful purple hue over my surroundings. At that point the colour of the shadows was transformed into a green comparable in clarity to a sea green and in beauty to an emerald green. The effect grew ever more vivid; it was as if we found ourselves in a fairy world for everything had clothed itself in these two lovely colours so beautifully harmonious with one another. When the sun had set, the magnificent display finally faded into grey twilight and then into a clear moonlit night filled with stars.'

Red projector only.

Vermilion and white produce green

Projecting vermilion alone produces vermilion only and where shadows fall they are dark. But when the shadow is lightened by white light it turns green. This appears to be inexplicable since only one colour is being projected. Nevertheless, the eye appears to seek a complementary balance.

When the coloured shadows are isolated they turn pale grey. Nevertheless, as our examples show, the coloured shadows can be photographed. So the neighbouring colour is necessary. Our eyes evidently tend to create a whole, and the coloured shadows cannot be explained by the wavelength of the colours.

Green and white produce red.

Green and white produce red

When green is then projected, red appears in the lightened shadow.

Additions

When several colours are projected together new unexpected colours emerge.

Vermilion and green produce yellow.

Vermilion and green produce yellow

Vermilion and green shadows can be seen in the picture. The addition produces yellow.

Violet-blue and green produce cyan

The violet-blue and green (here light green) shadows are visible in the picture. When the two colours are added the result is pale blue (pure blue), as seen in the child's face.

Vermilion, green and violet-blue produce pale grey (white).

Vermilion and violet-blue produce a shade of magenta

Here we see a vermilion and a violet shadow. When the two colours are added the result is a wonderful magenta (pure red).

Vermilion, green and violet-blue produce pale grey (white)

All three colours combined are neutralized to make a pale grey which may be regarded as white.

Violet-blue and green produce cyan.

Vermilion and violet-blue produce a shade of magenta.

Here we see all the colours of the colour circle. Everything is balanced out by the complementary dynamic of the coloured shadows and by the additions.

Here all the colours of the colour circle unite: magenta, vermilion, yellow, green, pale blue, violet, white and black.

Coloured shadow theatre

We are all familiar with black and white shadow theatre, but playing with coloured shadows is much more interesting. This can involve either living actors and/or objects of any kind (made of cardboard, fabrics, coloured foils etc.) that cast shadows.

Use a white projector combined with colour projectors to create the scenery, and generate atmospheres with the coloured foils.

After-image colours

After-image colours, also known as complementary colours, are created by the retina when you look steadily at the same colour for at least a minute. What might be described as small 'wounds' are inflicted on the retina which the eye heals by means of the complementary colour. The colour can be projected either on to the eyelid or on to an external surface.

If you look for one minute at this picture and then transfer your glance to a white surface you will see the same image in its complementary colours.

Rebirth of the phoenix

For the 'Museum at Night' festival in Berne in 2007 (see p. 8) the image of a phoenix was painted on to the screen of the rainbow instrument. Observers were told to stare for one minute at the middle of the figure and then transfer their glance to the blank circle next to it. There the legendary bird was reborn in its complementary colours.

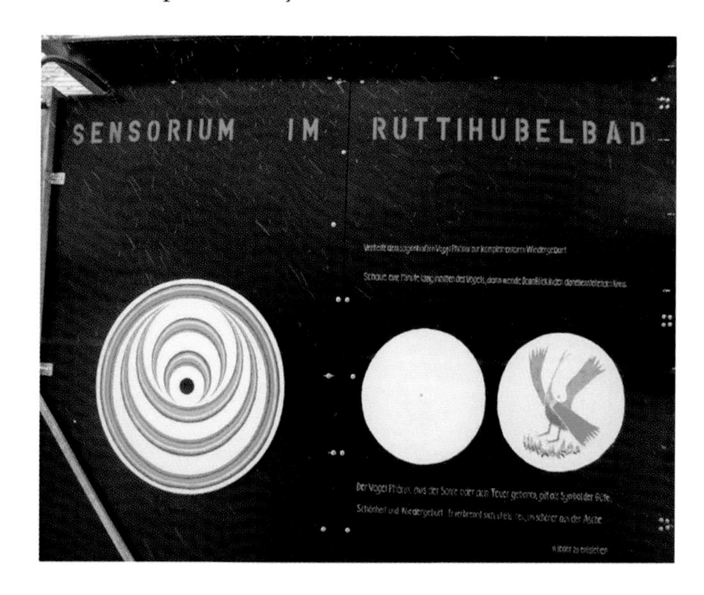

Born of the sun or of fire, the phoenix is regarded as a symbol of benevolence, beauty and rebirth. It is for ever consumed by fire from which it reappears out of the ashes with even greater beauty.

Soap bubbles

Children and adults alike have always been fascinated and delighted by soap bubbles. The fragile spheres or, when deposited on a smooth surface, hemispheres mysteriously produce colours and/or reflect their surroundings. They are one of the simplest and yet at the same time most complex phenomena – simple because the soap mixture can be obtained at a small cost in any department store, and complex on account of the interference colours which arise, because the bubbles are in fact double spheres with a minute space between the two layers, so that they refract the light just as do the raindrops in a rainbow.

True soap-bubble enthusiasts will no doubt be satisfied with the phenomenon itself as a sufficient explanation.

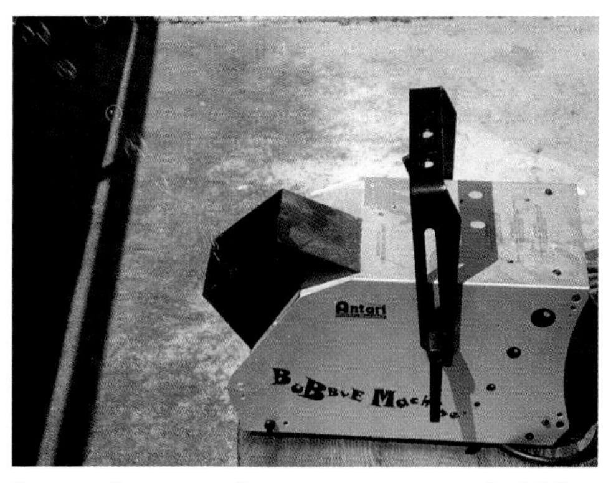

Some shops stock easy-to-use soap bubble machines.

Every soap bubble has a life of its own, and no two have identical colours. You can tell by their colour when they are about to burst.

Soap bubbles from the department store

One can study the brief life of soap bubbles using a bubble mixture from any shop. They are very colourful, mostly green and magenta at first, then yellow and finally transparent just before they burst. Sun and sky are always reflected in them.

Bubbles are in a state of flux. The soapy water flows downwards from the upper half of the sphere which gets thinner and thinner until it bursts.

Soap bubbles are tremendous fun and so simple to produce.

Soap bubbles shimmer in a variety of colours, not only revealing their inner state but also reflecting the world around them.

Experimenting with soap bubbles and observing them

If you deposit a soap bubble on to a pane of opaque glass or a wire frame, it is easy to observe and photograph when the light is right. The bubble has to reflect the light shining on it from outside.

The lighting is important when photographing soap bubbles. Here a PVC pipe containing a halogen lamp (12V/20W) is fixed to shine down on to the bubble. The bubble is then deposited on to the pane of opaque glass.

The bubble can also be stretched on a wire frame to make its shape square. The constantly changing and flowing layers of colour generally appear green and magenta. The top of the bubble usually turns yellow and then becomes transparent as it gets thinner before bursting.

The lower half of the bubble is the reflection of the upper half. The green and magenta stripes are typical. These are interference colours which are caused by the specific and constantly changing optics of the double bubble. We shall later also show how the green and magenta colours come about in connection with the edge colours.

DARKNESS AND LIGHT

Darkness is the mother of all things, the primordial foundation of all being and becoming. Out of the primordial foundation light begets the world of external reality, the visible world.

Darkness

We are seated in a room that is as dark as possible. We were led here with our eyes closed and we do not know how large the room is. The chairs are arranged in a circle and by holding hands with one another we each find a place to sit down. We let go of one another's hands; we sit there and we see that we cannot see anything at all. Or can we?

Opening or closing our eyes makes no difference. Yet something flickers. There are threads, whorls, white, grey, coloured. Each of us sees something a little different. After a while the flickering settles down. We begin to talk with one another. We have all had different experiences with darkness – now and in earlier times: 'Darkness scares me; it gives me shelter; it's absorbent; it generates warmth. Darkness enables me to experience my environment right up close.' When others speak we find ourselves inside their voice and sense their very being from within. All our senses are more active: listening, smelling, touching, our sense of temperature, our concentration on the thoughts of others and also on our own thoughts. The darkness is the mother of all things, the primordial foundation of all being and becoming.

Light

A beam of light pierces the darkness. That is what we think happens. And yet all we can see is the source of the light and the spot it illuminates. We see something where the light comes up against matter and is reflected back – but what is it we see? We see the object, that which is across from us, the object upon which the light is shining.

Light is invisible, so no one has ever seen it. It remains invisible until it falls on a physical object. It moves at a specific velocity, but we still do not know whether this takes place in the form of particles or waves, or both or neither. Light remains a mystery, as does darkness. Both are invisible.

Light has something of the straight line about it. It causes reflections and turns the world into a counter-world, an object. Out of the primordial foundation, light begets the world of external reality, the visible world.

The journey from darkness into light

Now we bring light into that dark room with a candle. How wonderful it is, this candlelight! Characteristically it does not entirely banish the darkness. We experience the partially illumined room in a dreamlike way as something that exists between the motherly darkness and the fatherly centre of light. We begin to think about the journey human beings have undertaken. They too set out from the warm and magical darkness and have now arrived in the cold world of objects in which everything is fully illumined and reflected. While early humans celebrated their mystery rites in darkness – as witness cave paintings – nowadays even the night is made bright with lights. All that is dark and shadowy has to be banished by reflective light. Rationality, the cause and effect of scientific knowledge, has turned the world into a brightly-lit and alienated counter-world.

Between the primordial beginnings of humanity and the present day, human beings experienced, and still experience, the dreamlike, mythological world in which everything was still whole. Darkness and light complemented one another; that which was inside was also outside; and as above so also was below (Trismegistos). Humans made images of the world, fairy tales and myths came into being. The formula (below) indicates the journey experienced by humanity:

Early times
Darkness
Magical

Intervening ages
Candle in the dark
Mythological

Modern times
Fully illumined room
Reflecting (mentally)

While the dark, magical world is linear, the mythological world is pictorial; and the post-Renaissance fully-illumined, wholly perspective world encompasses all space.

One cannot help asking what will

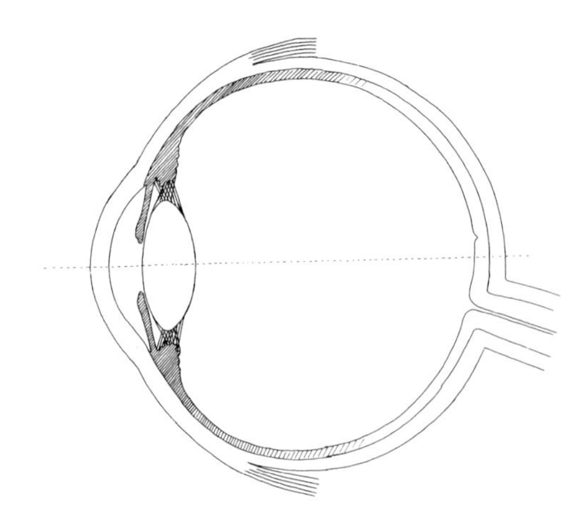

The structure of the eye (after: Lothar Vogel, *Der dreigliedrige Mensch*, 1979).

A candle flame between light and darkness displays a wonderfully coloured aura.

happen next. We have arrived, so to speak, at the end of the road, the threshold of consciousness, where we have finished up in a cul-de-sac. The present state of human society makes this obvious. What we need is a leap of consciousness across the threshold, up into a new level of awareness. Many people are seeking to cross this threshold by means of drugs or violence, through the world of virtual reality, by altering DNA or along various esoteric paths. The path of colour leads us from the magical to the mythological world and finally to the realm of perspective space. But can it take us further?

Since the beginning of the last century artists have been endeavouring to move on from the world of impressions (the outer world) to that of expression (the inner world). As a magician of colours and an integrator of light and darkness, Vincent van Gogh has shown us in his work how we might proceed (see p.92). Others, such as Braque, Picasso, Klee and Kandinsky, sought to master the external world of space in Cubism and abstract art. And psychologists have discovered (or rediscovered) the dark domain of the unconscious.

A good while earlier Johann Wolfgang von Goethe had corrected Isaac Newton's colour theory by demonstrating that colours arise not only out of light but also out of darkness. And now post-modern humanity is working on integration. So-called evil and darkness must not be rejected or fought-against as something outside one's own existence, but ought instead to be accepted as a part of oneself. Our shadow complements the brighter part of our nature. How can light and darkness, both forces of creation, come to form a totality? Through colours which arise not only out of light but also out of the darkness!

Goethe's theory of colour

Goethe published his theory of colour in 1810 when he was 61 years old.

The six chapters begin with the physiological colours which originate

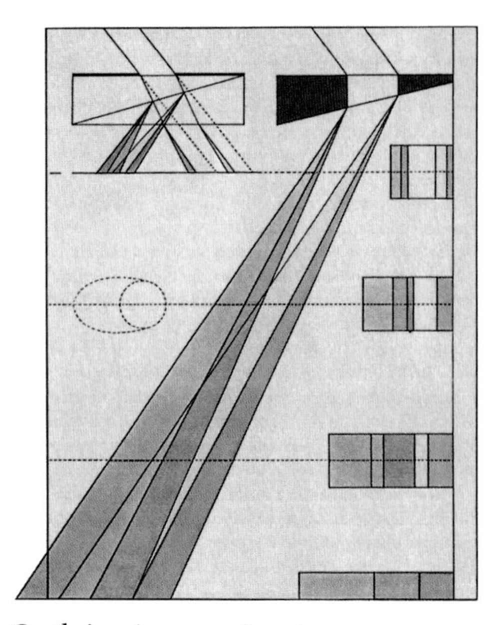

Goethe's prism experiments (from: J. W. Goethe, *The Collected Works, Volume 12, Scientific Studies*, 1995).

as complementary colours inside the eye. Then come the physical colours (light – darkness – turbidity; prismatic colours), the chemical colours (e.g. acidic and alkaline reactions), the chapters 'General Observations on the Principles of Colour' and 'Relationship to Other Fields'. And finally we reach the chapter 'The Sensory-Moral Effect of Colour' in which the effect of colours on man's inner nature is described. By following this journey from so-called subjective colours, i.e. those that are seen and generated by the human eye itself, through to the colours in the worlds of physics and chemistry, and then back again to the sensory-moral effect of colours upon the soul, we have completed the circle of a science which includes rather than excludes the human being.

Goethe himself considered his theory of colour to be his most important work. Why? Because in its examples it shows what a true science ought to look like if it combines complementary opposites, i.e. what the senses perceive with the meaning we ascribe to those perceptions.

Goethe's theory of colour contains descriptions of a thousand perceptions. He did not theorize, he gave descriptions and in doing so he organized his perceptions into the abovementioned chapters. And, contrary to Newton, he concluded astonishingly that colours arise not only out of light but also out of darkness. The theory of colour could serve as an example of what a future sensory-moral science might look like also in the realm of the other senses, i.e. that of touch, life, movement, balance, smell, taste, temperature, hearing, speech, thinking and the ego as hinted at by Rudolf Steiner in his own theory of the senses.

Goethe's theory of colour is little known today and certainly not accorded recognition by the scientific community. Why? Because its point of departure remains contrary to present-day scientific practices in which the human being is still not trusted to be a proper observer; so observations must be made by instruments independently of human beings. People call this being objective. And yet the things which every one of us sees every day with our own eyes are subjective in the good sense if we train our organs accordingly.

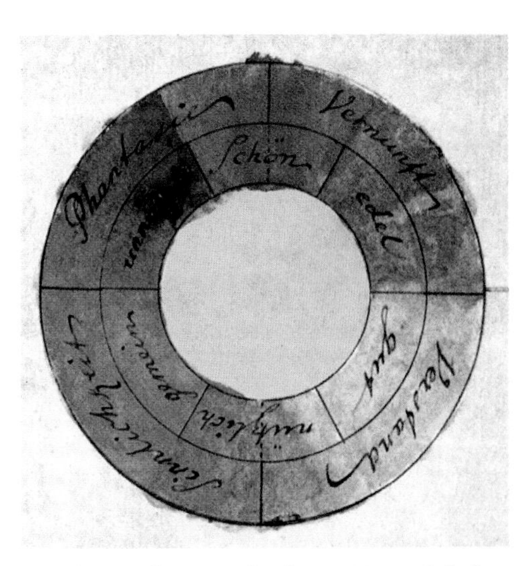

Goethe's colour circle (from: Hans Gekeler, *Taschenbuch der Farbe*, 1991).

Paradoxically, the theory of colour is scarcely known today even among those who admire Goethe. Merely reading books on the subject is not sufficient for a proper understanding of it. The colours themselves need to be seen and observed, day in, day out, as long as the eyes can see. It is surprising that in today's world, which tends to be so very visual, coloured shadows (the phenomenon of simultaneous contrasts), for example, are virtually unknown although they surround us every day. Or that magenta, well-

Sunlight is dimmed by the gloom and becomes yellow-vermilion.

known technically in the printing industry, is little understood as the pure red it really is. What we generally call red is actually a yellow-red. Goethe's theory of colour has to be experienced. Colour can only become an identity when we take it as evidence of what we see, as the truth of what we perceive.

We want to see the truth of colour over and over again, for example the colour of the sunset (darkened lightness), because we can identify with the primordial phenomenon. I and the world communicate here in a higher sense. A spiritual rite takes place upon the altar of nature, just like that conducted by Goethe when he was a child. This is communion not in a symbolic sense but as a form of identification. I and the world, the world and I become one. Nothing is left in the background, for the internal becomes the external and the external becomes the internal.

In the third century after Christ, Manichaeism developed a cosmogony of light and darkness which became the foundation for Goethe's theory of colour.

Manichaean cosmogony

'The myth of light and darkness is rooted in the teachings of Mani (AD 215–276), founder of the Gnostic religion of Manichaeism. The ideas of this spiritual stream reappeared later in the Bogomil movement in Bulgaria (eighth century), in the Cathar movement in southern France (twelfth century), and among the Templars, the Rosicrucians and the alchemists. Goethe also moved within this tradition.

'Mani's view of the world has ceased to bear the serene, bright stamp of the Olympian realm. The gods no longer descend to the human level; instead human beings rise up to become spiritual beings devoted to divinity who leave the world behind them and enter into the spheres of light. Mani himself, however, remained attached to our world. The diurnal miracle of the way in which sun and moon travel in the firmament was an expression of hidden forces and powers which revealed themselves to him as the spiritual motivators and bearers of the world. He saw the external processes as stirrings of an ensouled world body; sun, moon and stars served as images which ... reminded him of the spiritual origins of the earth.

'Cosmic architecture is determined by the number twelve. The circle of the sky is occupied by the twelve great gods, three in each quarter of the heavens. So the One around whom they move is described by the Manichaeans as the "fourfold Father of Greatness". They also saw the order on the earth as being in accord with the cosmic order, so they arranged the constitution of their church to be in keeping with this. Thus Mani surrounded himself with twelve Teachers who corresponded to the twelve force centres of the Zodiac.'

From: Eugen Roll, *Mani der Gesandte des Lichts*, 1976

The seven metamorphoses of the earth

The number twelve is the fundamental number of space, so whatever takes on a shape in space is subject to it. By contrast, what advances in time does so in accordance with the number seven. Just as in *Genesis*, the First Book of Moses, the process of creation follows the seven-day rhythm, so are almost all cosmic developments based on the number seven. Rudolf Steiner described the seven phases of world evolution in detail, giving them the designations Old Saturn, Old Sun, Old Moon, Earth, Jupiter, Venus and Vulcan. The evolution of the world according to the Manichaeans follows a similar course, as described in their literature. The most important indication came from Mani himself when he reminded his king: 'You know that of Earth there are seven!' In saying this he stated clearly that the earth will proceed through seven cosmic ages (aeons). The diagram below will illustrate this.

The scheme shows the seven cosmic metamorphoses of the 'earth' in such a

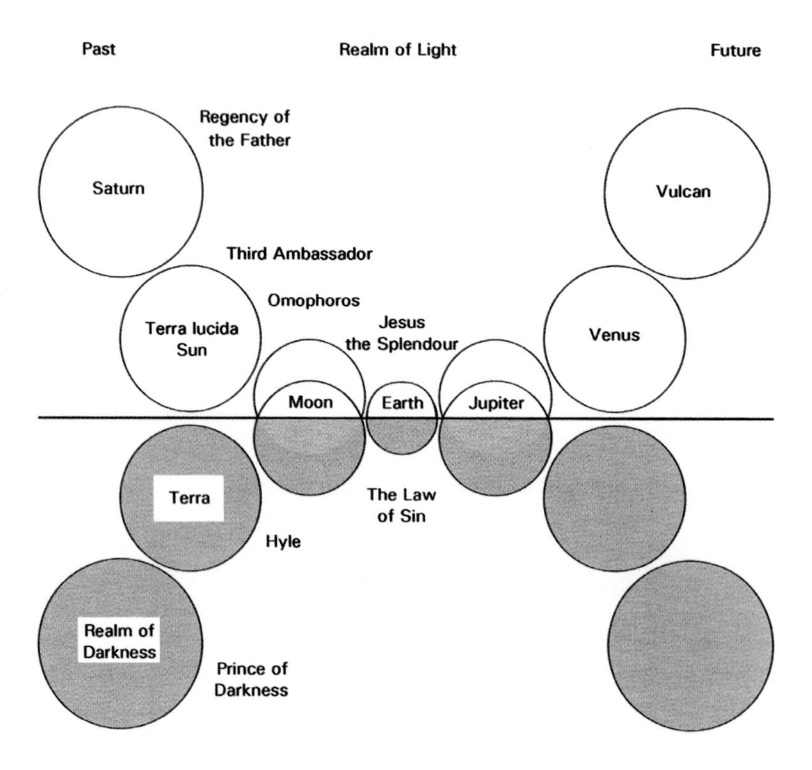

The seven metamorphoses of the earth (from: Eugen Roll, *Mani, der Gesandte des Lichts*, 1976).

way that they appear to be mirrored down below in a darker medium. The horizontal line dividing the world of light from that of darkness may be imagined as a darkened mirror resembling a black onyx which absorbs most of the projected light. The descent from top left down to the middle indicates a process of densification and thus of darkening. The ascent of the mirrored images in the lower left half suggests a process that as it were brings about a kind of lightening. From the Manichaean system we know of the 'terra lucida' and its opposite, the 'terra pestifera'. And the *Kephalaia* tells us of the regents of earth's planetary metamorphoses. In the fourth chapter Mani writes of the four great days emerging one out of the other. According to this the first, the Saturn day, is ruled by the Father (the 'god of truth'). The second, the Sun day ('terra lucida') is guided by the 'third ambassador' who lives in the ship of light. And we are to become personally acquainted with the third day.

While the left-hand side of the diagram reaches back into primordial times, the right-hand side tells of the far distant future of the earth. What we know as the actual 'earth condition' of our planet is shown at the centre of the picture in the way it was experienced by the Manichaeans. Here the mingling of light and dark elements has reached a stage which amounts to a kind of balance. The regent of this 'earth' is the one to whom Mani gave the title 'Jesus the Splendour'.

Spear and chalice

Even in megalithic times (around 3000 BC), light and darkness as the primordial forces of life were represented as a sunlike shining emanation and a protectively cavelike enfolding.

The rays of the sun, active in their 'manliness', appeared as an axe and the feminine inwardness as a moonlike square; Helios the spear-hurler and the archetypal mother as the waxing and waning moon. The light begetter and the dark birth-bringer became realistic symbols of life.

The Manichaean tradition gave us the legend of the grail which in the spear and the chalice describes the highest values of humanity. In the grail ceremony of Wolfram von Eschenbach's *Parzival* they are depicted as the blood-stained spear and the life-giving grail. Here the spear is the instrument of Christ's death on the cross while the chalice, which has gathered up Christ's blood, maintains the life of the grail community in the spirit. The sun element becomes the lance of death, bringing enlightened knowledge, while the moonlike chalice depicts the force of life forever renewed. That which is light becomes like hell (Jean Gebser) and that which is dark awakens life. The two together complement one another.

There was no contradiction in beholding the sun at midnight, for it was a spiritual reality.

LIGHTENED DARKNESS – DARKENED LIGHT

Darkness itself is invisible. To the human eye it only becomes visible when it manifests in violet and blue. The darkness 'shines' through the violet and the blue. They are the colours of darkness. So the process leading from light into darkness arises through an increase of turbidity.

Violet – blue – white

We are seated in a dark room where the light of one candle is shining. We have heard that although light and darkness are the creative forces they are actually invisible. So where do they become visible for our eyes?

Each of us in the circle now holds a candle at eye level and examines the flame. The candle-flame itself is an image of wholeness, an organon. The hard white wax melts and is sucked up by the carbonizing black wick where it burns in the form of gas: the colours blue, yellow, vermilion and violet arise. Not only is the flame itself coloured, for it is also surrounded by a coloured aura. The flame is like an alchemist transmuting one state into another.

Looking more closely at the blue colour in the candle-flame we find that it is deep blue when the background is dark but pale blue to transparent if we hold our hand or another candle behind it. The blue arises out of darkness; lightened darkness becomes blue. There are many examples of this in nature: the dark depths of a lake are lightened by the water, so that blue arises; by day the absolute darkness of the cosmos is lightened by the earth's atmosphere; the blue of the sky grows deeper and more intense the higher we climb up a mountain; astronauts fly

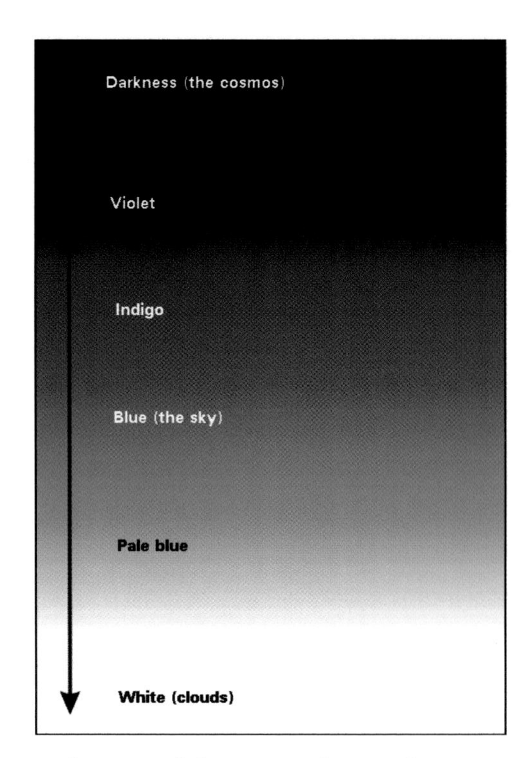

Lightening of the cosmos brings about violet – blue – white.

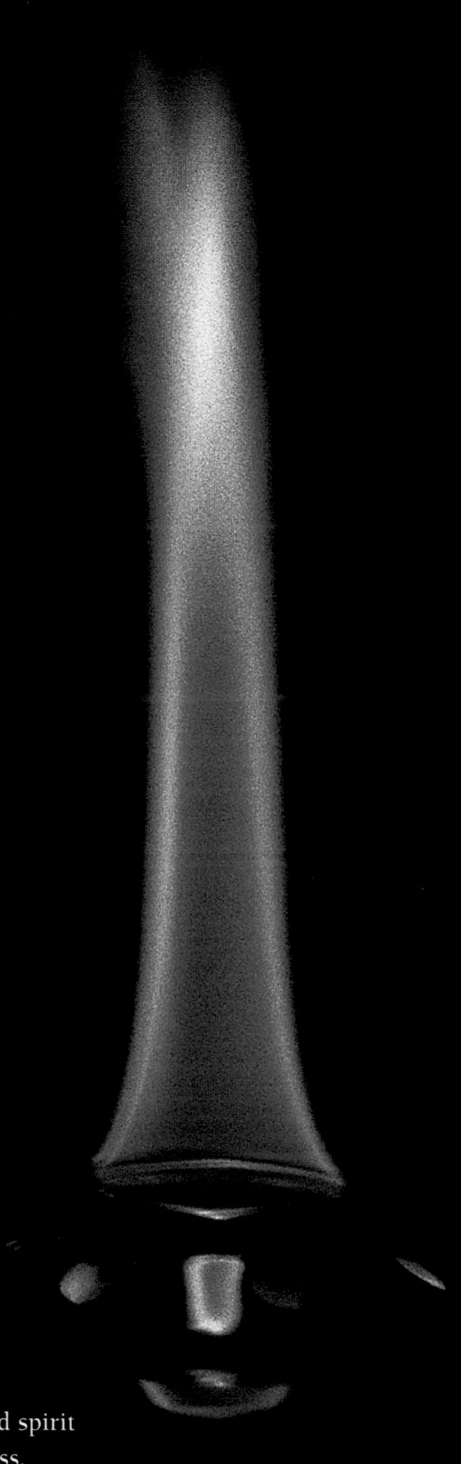

The blue of a methylated spirit
flame: lightened darkness.

first through the deepest indigo and then, before arriving in the utter darkness of the cosmos, the colour they see is violet. The more the atmosphere is filled with humidity, the lighter does the sky become – pale blue or even white – when we look out into the cosmic darkness from earth. When high humidity lightens the dark background to blue, the landscape also appears blue. But it appears very close up with no blue when the wind is hot and dry (foehn).

In the colour process leading from darkness to an increasingly lightened darkness, the darkness itself is invisible. Only when the violet and blue appear does it become visible. The darkness shines through the violet and the blue. So these are the colours of darkness.

Yellow – vermilion – black

Bright light from a projector is filtered through white paper. A light white results. When we block the light source increasingly with more and more layers of paper the light is not merely increasingly darkened: we see pale yellow, then deep yellow, and finally orange to vermilion. Then even more layers of paper completely prevent the light from penetrating so that it becomes dark, indeed black. So darkened lightness leads to yellow, orange,

The blue of the methylated spirit flame is deep blue against the dark background and pale blue to transparent against the light background.

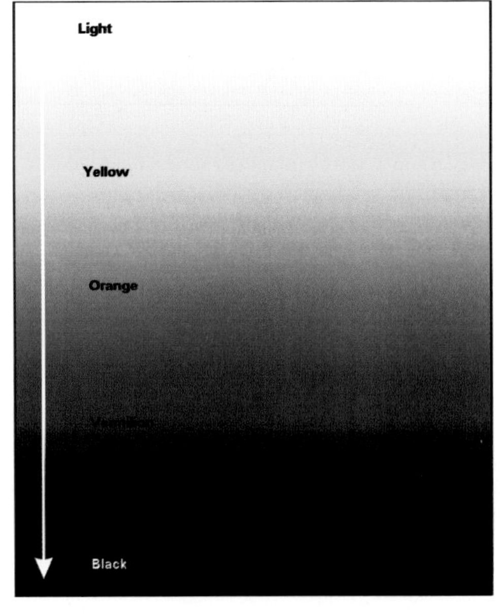

Yellow and yellow-red at sunset and sunrise.

Sunrise between the Schreckhorn and the Wetterhorn on 26 January 1995 at 8.15 a.m. seen from Ins.

Darkening in the laboratory by increasing layers of white paper.

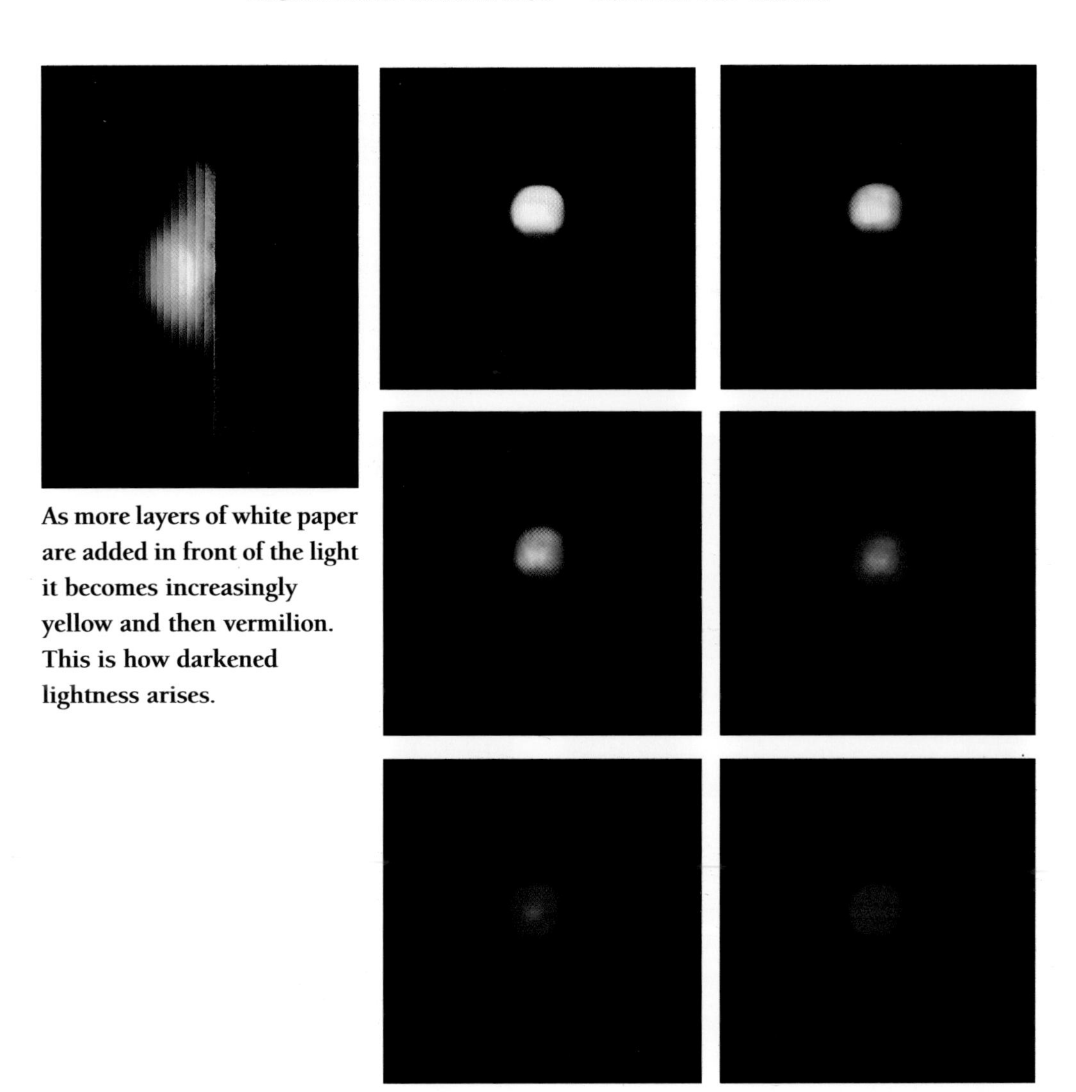

As more layers of white paper are added in front of the light it becomes increasingly yellow and then vermilion. This is how darkened lightness arises.

vermilion and black. The colour process can go no further than vermilion before disappearing into black. The same phenomenon arises at sunrise and sunset, when iron glows, when charcoal glimmers and when we look at the bright sun through glass blackened by soot. This colour process leading from light to darkness (black) arises through increasing turbidity brought about for example by layers of paper, atmospheric pollution, humidity etc.

A beam of light through turbid water

A beam is projected through clear water in a water-tank. The beam is scarcely affected. When soft soap is mixed into the clear water, and depending on whether the background is light or dark, the turbidity appears yellow and red or else blue – pale blue – white.

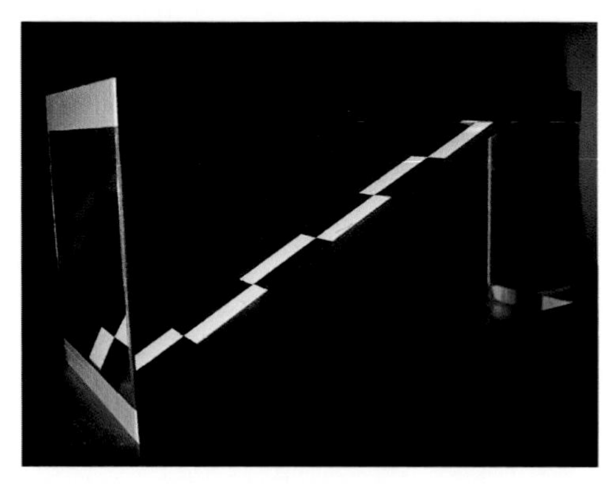

Clear water does not give rise to any colouring.

Light shone through turbid water is coloured.

Yellow-orange: darkened lightness.

Violet-blue: lightened darkness.

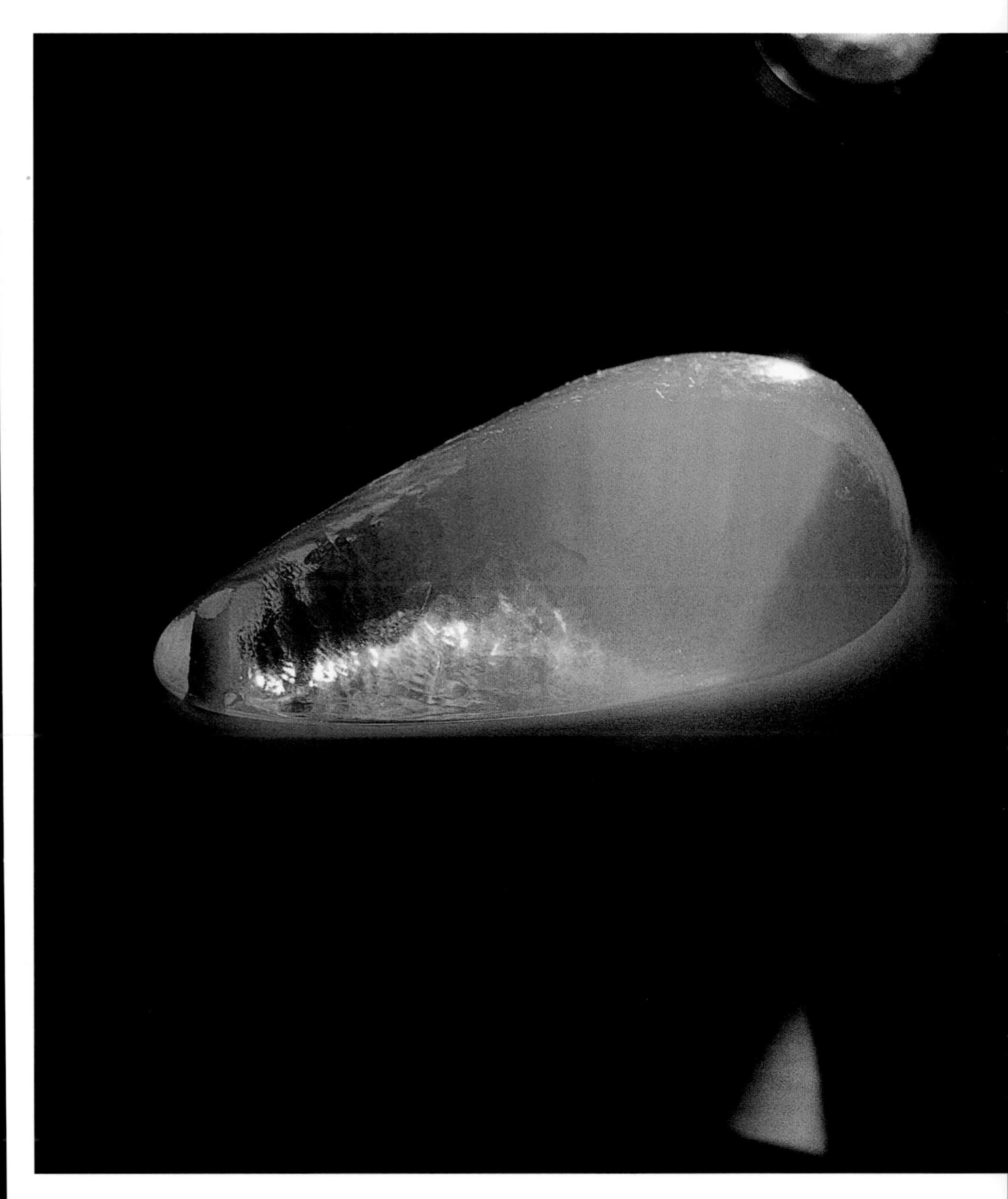

Turbid opaline glass displays a nuanced variety of colour, depending on whether the light is darkened or the darkness lightened.

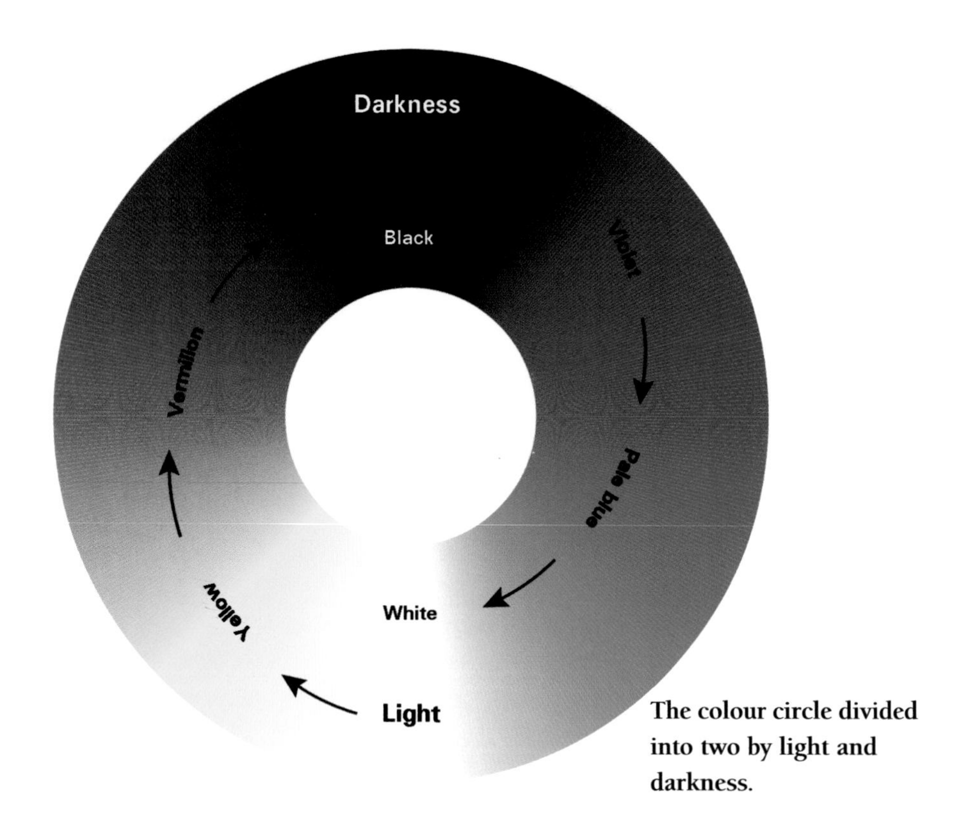

The colour circle divided into two by light and darkness.

The colour circle as a process through darkness and light

We now come to an initial colour circle. Light darkened by turbidity yields yellow, vermilion and black. Darkness lightened by turbidity yields violet, pale blue and white. We can regard this journey – travelling via the corresponding colours of darkness and lightness – from darkness (black) to light (white) and back to darkness as our initial colour organon.

It is a process that can be given form through painting, movement, music or speech. Course participants are asked to move in the realm of yellow and then of vermilion. Their gestures become more impetuous and emotional until finally becoming rigid in black. The rigidity then slowly relaxes in violet, and in blue the movements grow more soft and inward. Finally in white they become quiet and light-filled. This can then be repeated from the beginning.

The colour journey as a path through the crisis of darkness

In yellow, i.e. lightness which is only slightly darkened, we experience the purity and weightlessness of light. Scarcely darkened by matter, pale yellow expresses joy, serenity and cheerfulness. Orange becomes warmer, more earthy, more sensual. In vermilion we come to the passions of feeling, of the blood, the outburst of rage. Fiery vermilion is the most active of all the colours; it contains the greatest agitation and excitement, an orgiastic fusion.

The descent into darkness, the colour black, is abrupt: black is the final point of density; it is paralysis of all activity, lethargy, dark void, immobility. To linger in this black realm of death is tantamount to relinquishing all life, to suffering the crisis of hopelessness, or bearing pain for the sake of pain. Black lies side by side with vermilion, death beside life, absolute agitation beside the greatest lethargy.

As the darkness begins to lift we enter into violet. After crisis and sickness comes the colour of transformation, of transparency, of spirituality. Only someone who has passed through profound suffering can look into the world of spirit.

As the darkness continues to lighten into indigo and royal blue it embraces us more intimately, more soulfully: our inner world shines forth in the blue flower of Romanticism, in a yearning for far-away places and for the secrets of life.

A further lightening to the pale blue of the forget-me-not, that charming little flower, lifts our mood to cheerfulness.

And arriving in the white of clouds or snow we once again enter into the image of the spirit, the serene emptiness of the light.

So now a new process can begin …

The journey from light through the crisis of darkness which leads back to light (white).

COLOURED EDGES

A beam of light shining through a prism is refracted yielding wonderful colours.

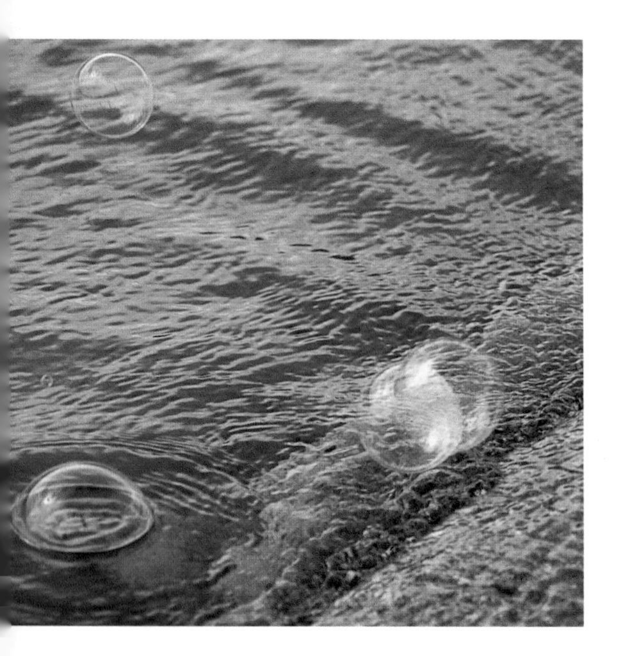

Soap bubbles as coloured objects.

Prisms

The prism is one of the most important instruments for the observation of coloured edges (yellow, vermilion, violet, blue) and for the production of green and magenta. The best ones are made from glass; cheap plastic ones are not so good. Large homemade water prisms are also impressive. Or sunlight can be reflected on to a projection screen by a mirror placed in a water-filled receptacle.

Phenomena such as the colours in dewdrops, in the oily surface of a puddle, in soap bubbles (magenta and green rings), in the technically produced iridescence of manufactured objects or in the aura around the moon or the sun are prismatic effects. In all these cases the colours change depending on the position of the observer.

Polished prisms of all sorts can be hung in the window; when the sun shines through them wonderful, shifting rainbow colours are projected on to the walls of the room.

A prism suspended in a window yields wonderful rainbow images when the sun shines.

Optical experiments using water prisms.

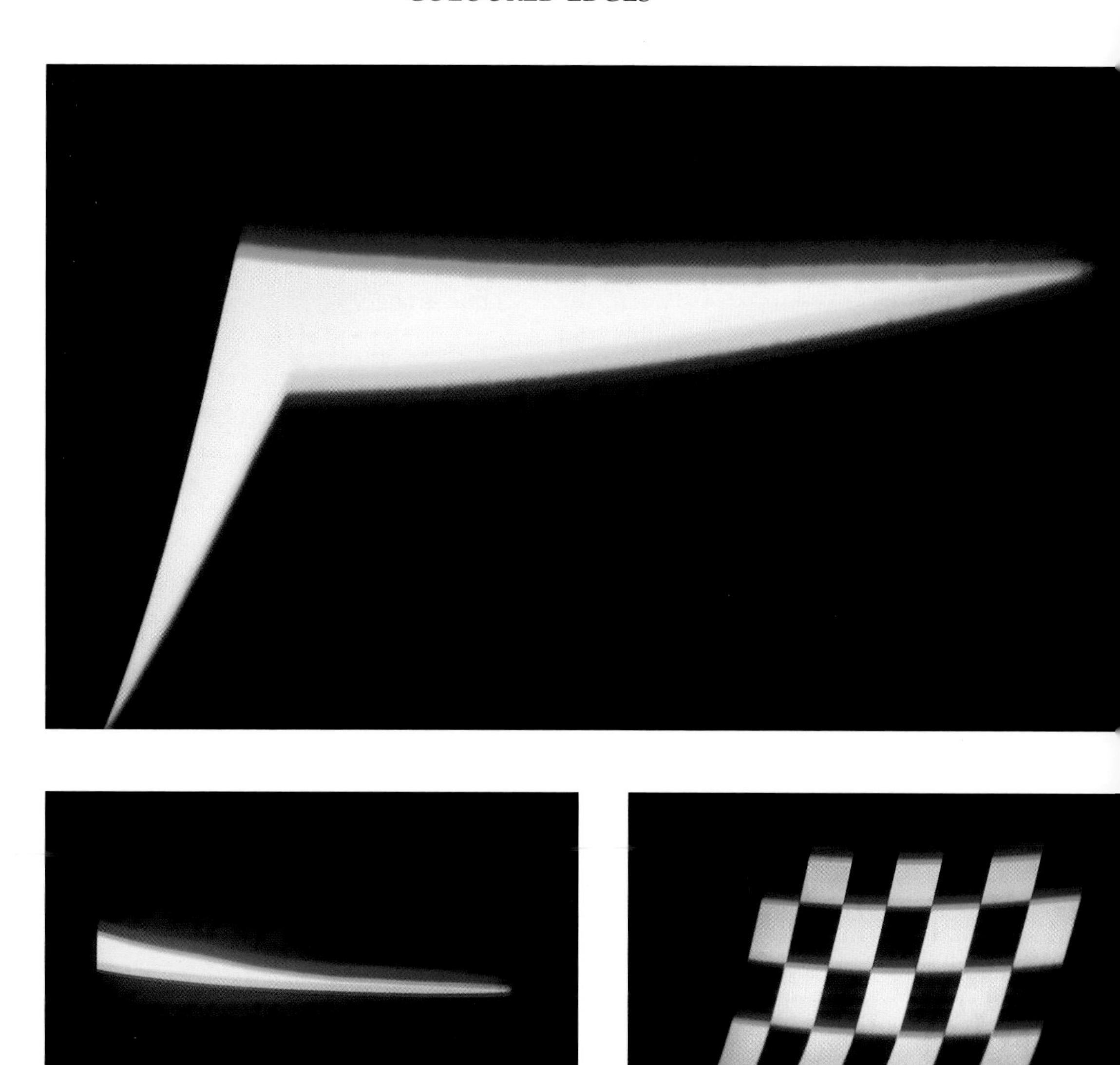

Black-and-white diapositive images projected through a water prism show colours at the edges.

Violet – blue – white

When we look through a prism at a black-and-white image, the lower part black and the upper part white, we see that the image is pushed downwards and the clear contrast between the black and the white is blurred. The upper light turbidity pushes down over the lower dark turbidity. The result is a lightened darkness. A small strip of lightened turbidity pushes down over the darkness, and a beautiful wide band of violet appears. Above this there is a narrow band of deep blue which dissolves upwards into pale blue and white. Here once again we have the same archetypal phenomenon as in the case of the blue of the candle, of the sky and the sea. The first lightening arising out of the darkness is perceived as a beautiful violet colour.

Observed with the naked eye.

Observed through a prism.

Yellow – vermilion – black

When we look through a prism at a white-and-black image, the lower part white and the upper part black, we see that the white image is pushed downwards and the clear contrast between the white and the black is blurred. The upper dark turbidity pushes down over the lower lightness. The result is a darkened light. Where only a small amount of dark turbidity is superimposed on the lightness, pale yellow appears; where the lightness is darkened more intensively we have orange and vermilion. The vermilion immediately ceases to glow in the darkness. The dark turbidity has transformed the lightness into black. The yellow-red edge is an image of sunset or sunrise, although here the evening or morning horizon is upside down.

Observed with the naked eye.

Observed through a prism.

GREEN

The interaction of the light and the darkness brings about the new colour, green. Green is the intermediary between light and darkness.

How green comes into being

Once again we are looking at a black-and-white image. A narrow white strip lies between black above and below. Again we push the image downwards by means of the prism. If we push it down just far enough to allow a small strip of white to remain between the coloured edges, we see the colours of darkened lightness (yellow, vermilion) above it, and the colours of lightened darkness (blue, violet) below it. But when the pale yellow above and the pale blue below merge together, a new colour appears: green. Here we have the darkening of the white strip.

So the colour processes arising out of light and those arising out of darkness are here united in the colour green. Green is the intermediary between light and darkness, and this is exactly what we see in the rainbow.

Observed with the naked eye.

Lesser refraction.

Greater refraction.

The rainbow

Let us observe the rainbow with the sun behind us and the shadow of our head in front of us. Inside the rainbow there is lightness, while outside it the sky is rather more dark. A deep vermilion followed by yellow emerges from the outer darkness. From the inner lightness violet and then blue emerge. Between them, uniting the two coloured worlds, is green.

According to the Bible, when God had had enough of humankind he drowned the whole world in the Flood. Only those who accompanied Noah in his Ark were saved on Mount Ararat. And then God made a new covenant with humanity, creating the rainbow as a sign of reconciliation.

A mood of reconciliation always arises in us when we observe a rainbow, whether through a prism, in a sunlit jet of water from a hosepipe in our garden, or against the sunlit curtain of rain after a thunderstorm.

And when we see it from a mountainside, from an aircraft or the top of a ladder, it manifests in a complete circle encompassing the shadow of our head.

And, by the way, each one of us sees our own personal rainbow (see below). The rainbow is the only imagination visible to our physical eyes.

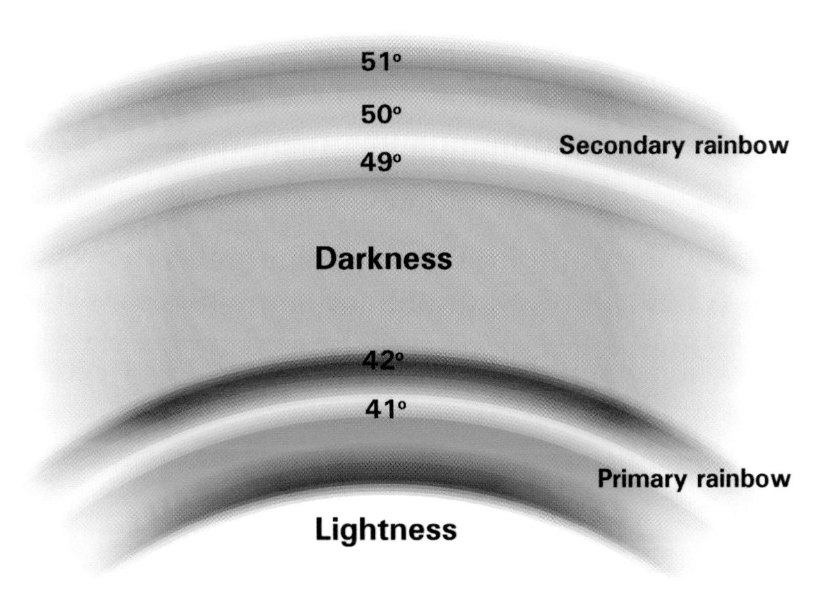

The rainbow is the only imagination visible to our physical eyes.

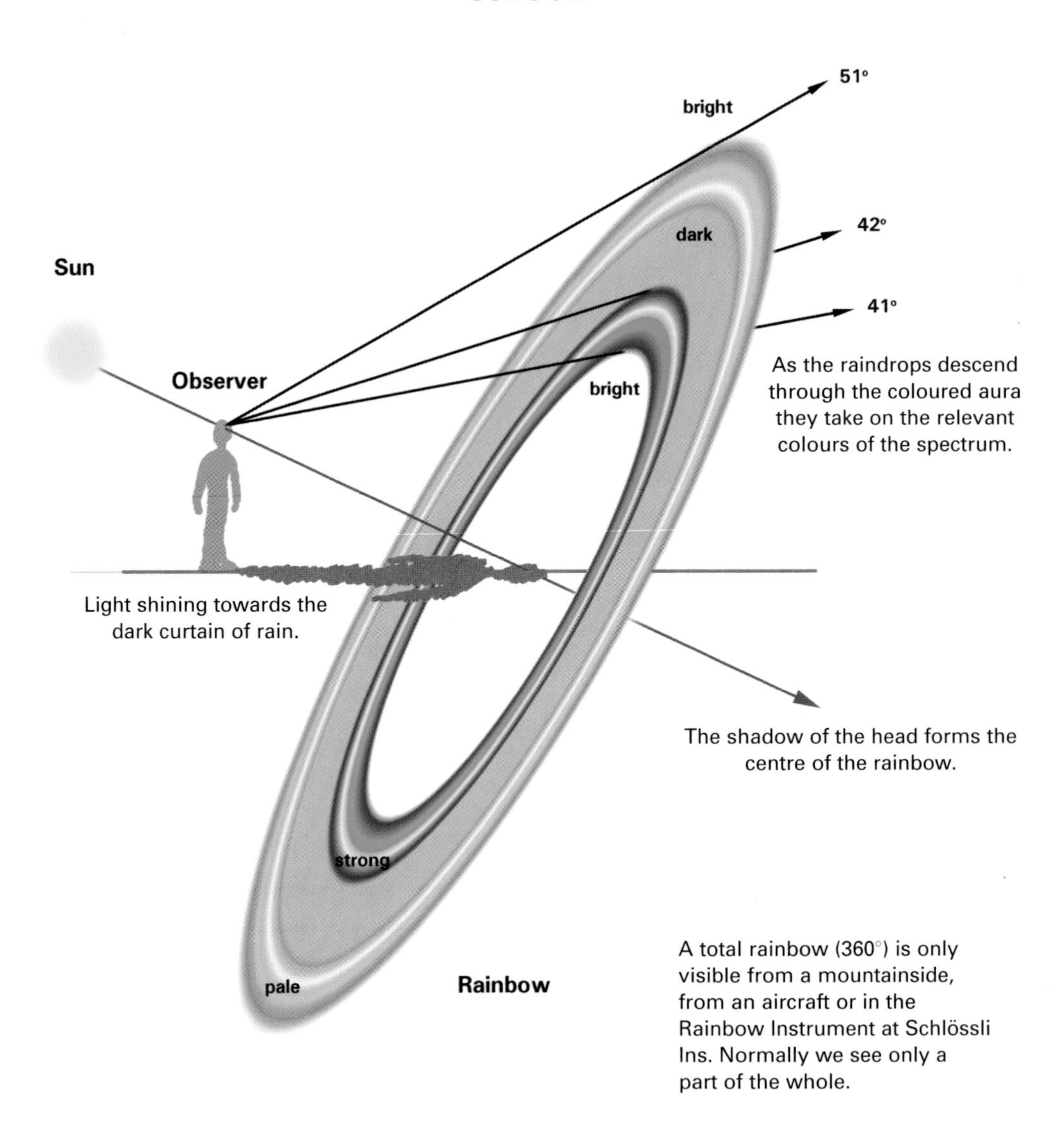

The Rainbow Instrument

A rainbow only arises when the sun shines straight at a curtain of rain. The shadow of our head is situated at the centre of our own personal rainbow. When the curtain of rain is far away the rainbow is huge. The closer the curtain the smaller the rainbow becomes, but also all the more perfect (more circular). Most of us have also seen partial rainbows in the spray of a waterfall, in a fountain or when watering the garden with a hosepipe. Based on all this, at Schlössli Ins we have constructed a Rainbow Instrument which, when conditions are right, gives a total rainbow, i.e. a complete circle (360°).

The Rainbow Instrument.

A horizontal pipe with nozzles that spray water vertically downwards is fixed to a rotating pole (250cm). When the sun is not too high in the sky (morning, evening, spring, autumn) you stand as close as possible to this curtain of artificial rain. Looking up at it you will see the complete circle of the rainbow to the right and left of the bottom of the rain curtain at about 45 degrees higher than your eyes. The shadow of your head will always be at the centre of the rainbow circle.

The pole can be turned to position the water-spraying pipe at right-angles to the direction of the sunlight. Here we have created the phenomenon artificially. But it can occasionally be observed in nature as the 'mountain spectre' you notice when standing on a mountainside looking down at a sea of mist below. Also when travelling by air you might see the shadow of the aircraft (the shadow of your head) rushing along on a carpet of clouds and surrounded by a complete rainbow circle.

The Rainbow Instrument.

The eye or the subjective objectivity of the rainbow

If you move about when observing a rainbow appearing, for example, in the spray of a waterfall, you notice that the rainbow moves with you. When you ask your companions where they see the rainbow, they each point to a different spot. So 'your' rainbow is seen subjectively by your own eyes.

The term 'subjective' is often taken to refer to things that are not objective, i.e. not generally valid scientifically. But this may or may not be the case. When you observe accurately where and how you see a rainbow you quickly arrive at objective, i.e. generally valid, laws. For example, you always see the shadow of your head in the centre of 'your own' rainbow. The angle between the axis sun/head/head-shadow and the primary rainbow which you see is approximately 42 arc degrees. The space within the arc of the rainbow is lighter, indeed almost white. The rainbow begins with violet on its inner edge, and this is followed outwards by blue, green, yellow and vermilion. The secondary rainbow is seen at about 51

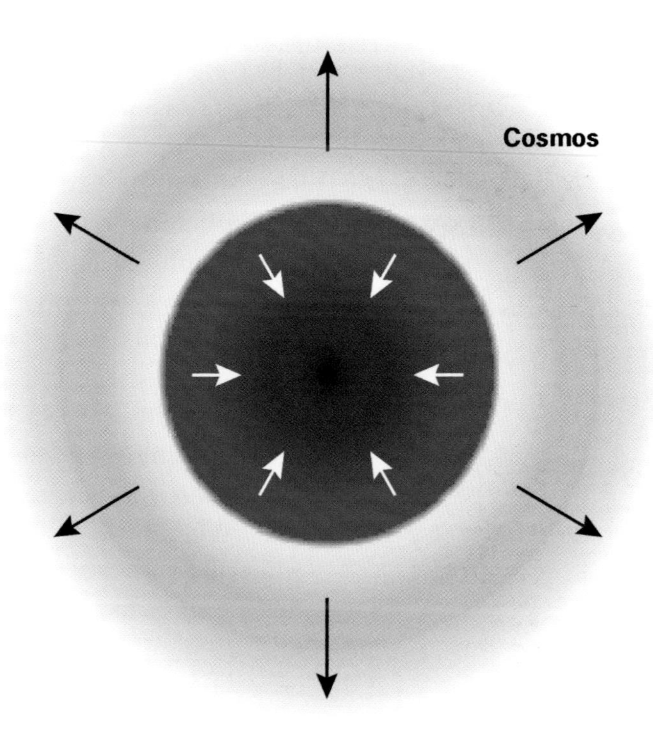

The plant covering of the earth between cosmos (sun, moon, stars) and the centre of the earth (mineral kingdom).

arc degrees outside the visual axis, and the sequence of its colours is reversed.

So anyone observing these things is confronted with the same optical laws. And yet they can only be seen subjectively. No one can see a rainbow on behalf of someone else. This is the subjective objectivity.

When you bathe in the ocean at sundown you find yourself swimming in a wonderful pathway of the sun's light. And if you look away from the sun and towards the breaking waves on the shore, again you momentarily see this sunny pathway. The sun approaches you and invites you to bathe in its beams. This union of the sun with oneself is deeply moving. Indeed, one is tempted to say, 'Not I, but the Sun in me' – a profoundly mystical experience.

TABULA SMARAGDINA HERMETIS.

VERBA SECRETORUM HERMETIS.

Depiction of the 'Esoteric Symbols of the Rosicrucians' (in Viktor Stracke: *Das Geistgebäude der Rosenkreuzer*, 1993).

Your love for the sun becomes infinite and yet you do not lose your own self. Thus your love for nature becomes love for the Godhead, for objectivity. And this objectivity is accessible to anyone who cares to bathe in the ocean at sundown. Every individual bathes in his or her very own subjective sun path. And if you have a loving heart you are amazed and deeply touched in the depths of your being by 'your' sun which objectively prepares a personal path for every individual.

Travelling by train at night you look out of the window and all of a sudden espy a pair of eyes looking at you, a mirror image of a fellow traveller. You see the face at the very point from which the other person sees you, subjectively – objectively, in accordance with the optical laws of the mirror. This, in turn, is the blending of I and You, of I and the World.

This subjective objectivity tells us that our eyes, which have been taught how to function as organs, are capable of taking a stance towards the world which enables them to unite with and yet remain independent of it.

It is our I which sees through our eyes. When our I is filled with love it will unite with and communicate with the world. As Albert Steffen put it in his diaries, we pour '... our I into all the laws of nature. Minerals, plants and animals attain I-consciousness through us, as do the stars and other worlds. We pass from one subtle I-experience

The plant world is fundamentally green, yet it blossoms forth in all colours.

to another. We should not cease in our efforts for we must work tirelessly to enable the whole world to attain its I. Within our being we must experience the whole I-being of the cosmos'. (*Das Goetheanum*, 50/1999) Here we have experiences of nature as mystical facts. We can tell others about these, yet all those who desire to participate in such mystical experiences must do so on their own account. A mystical experience, too, can quite concretely be regarded as subjective objectivity. It is an encounter with nature in which the I participates in discovering the world.

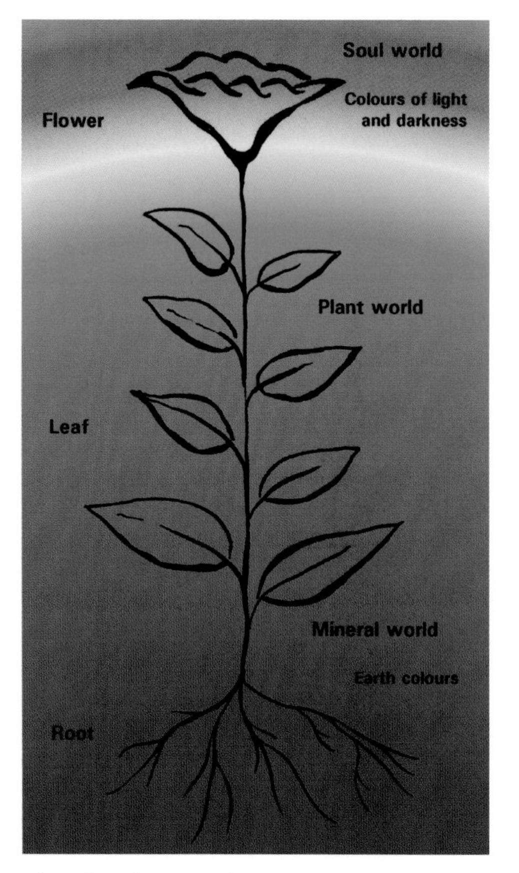

The plant between heaven and earth.

The green colour of plants

Why are plants green? Because they, too, mediate between two worlds: the dark world of the earth and the sunlit world of the sky. The sun and the centre of the earth are the two points of orientation of the geotropic (root) and heliotropic (stem) plant cells.

Red and yellow, i.e. the light colours, boost stem growth in plants. Blue and violet, i.e. the dark colours, delay stem growth and promote root development.

Green leaves with their surfaces mediate between the worlds of light and dark. In its vegetative zones the earth as a totality provides the green of the plants, which in the form of chlorophyll mediates between the centre of the earth and the cosmos in this process of light and dark.

And what about coloured flowers? In them the plant grows beyond its vegetative state and reaches up towards a higher inwardness of soul, as is shown by the shapes of its buds and flowers. The greenness of plants provides the vegetative basis for the soul colours of the flowers.

Green as an archetypal image of reconciliation

Hildegard of Bingen described green as the colour of Christ. On Easter morning Mary Magdalene saw the risen Christ as one concerned with greenness, a gardener. Christ as the Son of God brings reconciliation between God and human beings. So here again we have green as a mediator between two worlds.

The 'Tabula smaragdina', the emerald tablet of the alchemists, praises the principle of correspondences between microcosm and macrocosm.

The colour green has the greatest number of shades, ranging from the almost yellow of vegetation in spring via the rich green of meadows in summer to the deep green of pine trees in winter, and even beyond this to the ambivalent blue-green of the colour turquoise. Native inhabitants of the rain forests distinguish between hundreds of green shades in the world of plants.

As a region of rest, which at the same time also moves gently into the lightness of yellow and into the darkness of blue, green gives the colour circle its living, earthly pole.

In the spectrum, too, its place is in the middle between the long waves of vermilion and the short waves of violet.

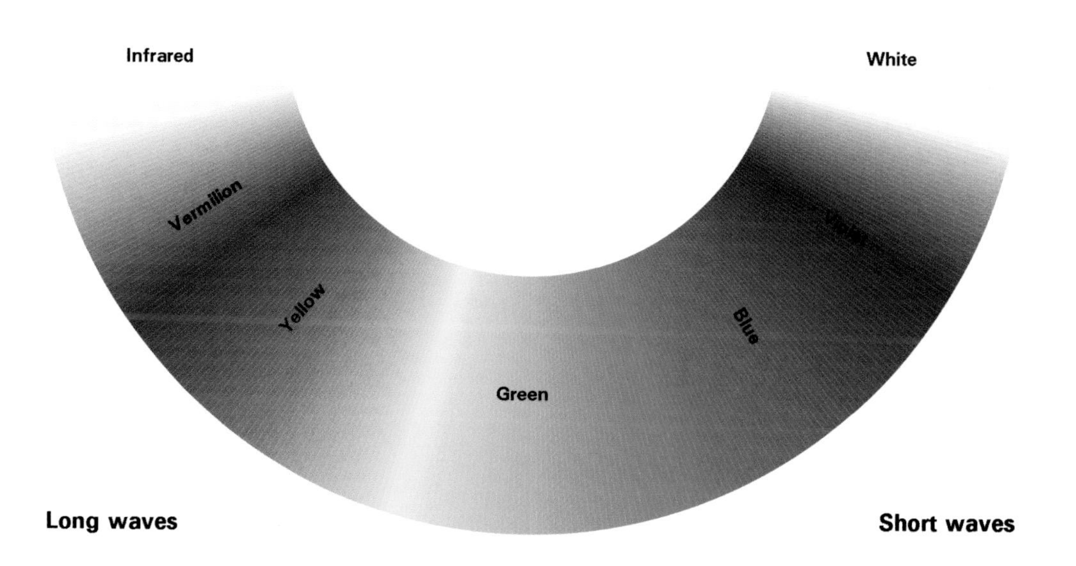

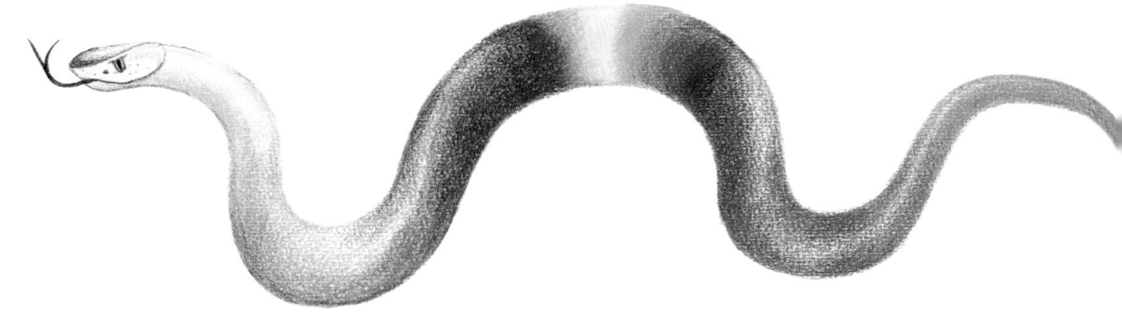

Radio waves Microwaves Infrared **Visible light** Ultraviolet X-rays Gamma rays

The electromagnetic spectrum

This consists of a continuum of wavelengths reaching from gamma rays to radio waves. Only a small portion of these energies – the seven-coloured spectrum of natural daylight – is visible to the eye. Like the rainbow snake, the electromagnetic spectrum is a profound metaphor for the unity between the comprehensible and the incomprehensible world. The rainbow snake is the first cosmological model for the spectral order of universal energy (from: Lawlor, *Am Anfang*).

Heat and light

One who enters into the world of sense via the Portal of Birth beholds

the Light.

At the end of earthly life, after death, the Fire of Purgatory awaits us

with Heat.

Bodo Hamprecht

Heat, like light, is not directly perceptible. Both are mediated by objects. Heat and light need physical matter in order to become manifest.

In the soul, heat arises through enthusiasm, love, anger, courage. Heat brings about movement and is a prerequisite for any human relationship. In his book *An Outline of Esoteric Science*, Rudolf Steiner described how heat is, in a sense, the primordial substance of creation (in 'Old Saturn'). Light came into being later, when the earth was recreated in 'Old Sun'. First heat, then light. The element of temperature also shows in the warmth of yellow and vermilion shades and in the coolness of violet and blue shades. In the soul, light signifies thinking, consciousness, clarity, enlightenment, truth.

The spectral colours reach from the heat of infrared via vermilion to the cold of violet and on into ultraviolet.

Wavelengths of the colours

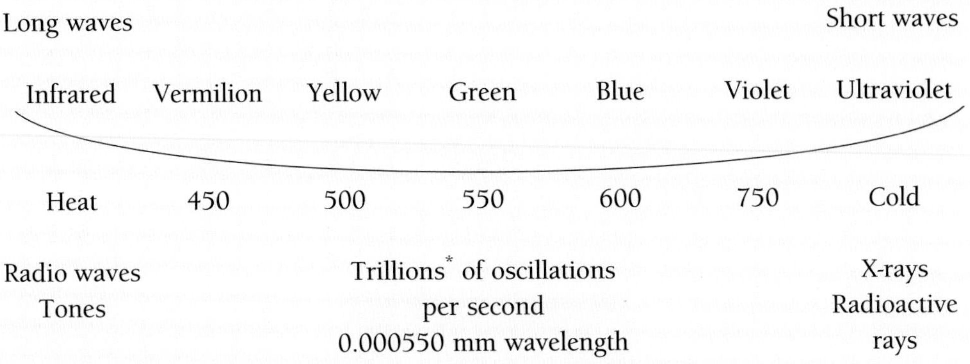

Long waves						Short waves
Infrared	Vermilion	Yellow	Green	Blue	Violet	Ultraviolet
Heat	450	500	550	600	750	Cold
Radio waves Tones			Trillions* of oscillations per second 0.000550 mm wavelength			X-rays Radioactive rays

Light: both waves and particles

* One trillion = one million million.

MAGENTA

Magenta is the royal colour in the hierarchical order of colours. Magenta lends the colour circle its spiritual pole. It is the colour which contains neither yellow nor blue and therefore also no green.

The origin of magenta

Once again we look at a black-and-white image, this time with a narrow black band between white above and below. With the help of a prism we refract the image downwards:

By refracting the image just enough to retain the black band between the coloured edges we manage to keep the light and the dark colour realms apart. But when the refraction is greater the black is replaced by colour. Magenta thus arises as lightening and intensifi-

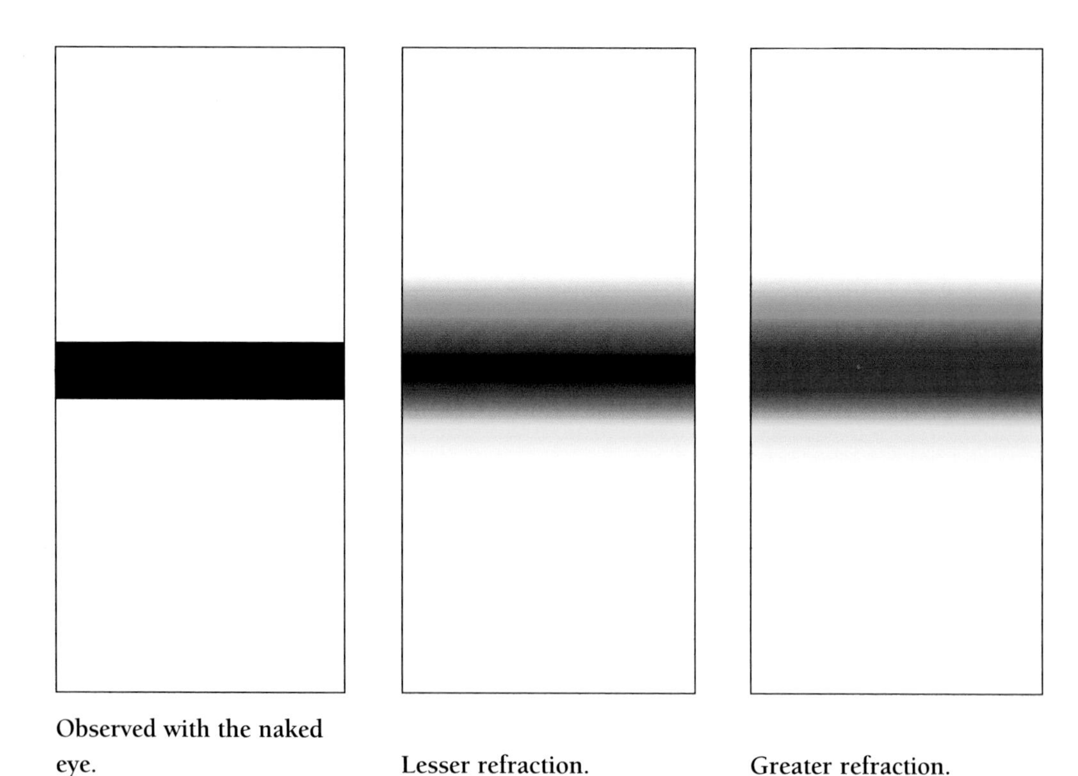

Observed with the naked eye.

Lesser refraction.

Greater refraction.

Looking westwards from Ins towards the Jura Mountains, the red colour of the evening sky is intensified right up to magenta.

cation. The edge colours of the spectrum, vermilion and violet, unite in an addition or intensification to form the royal colour magenta, also called rose, pink, peach blossom or the blushing flesh tint. Magenta crowns the colour circle as the opposite pole to the earth-based green. It often appears as a pale violet in the buds of spring. Just as green calms the processes of light and dark within the realm of light, in magenta we have a summary and intensification of these colour processes. Whereas green forms a stable, earth-bound pole, magenta as a spiritualized colour culminates upon a lonely, unstable summit. Mixed with yellow, magenta can descend towards a warm vermilion; or, mixed with blue, it brings forth shades of violet-blue.

Here magenta lies above the colours of the spectrum, so it is not a rainbow colour. But it brings about the completion of the colour circle which has thus far remained incomplete.

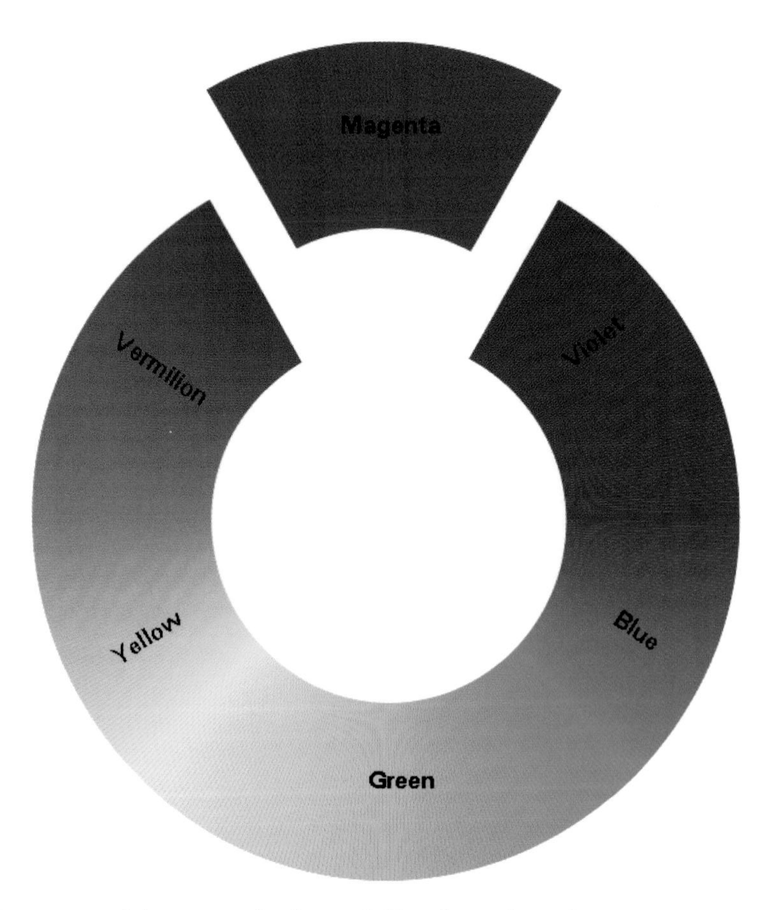

Magenta is not one of the spectral colours. It lies above the colours.

Green and magenta often appear adjacent to one another in nature.

Magenta, queen of the colours

In a hierarchical order of the colours, magenta may be regarded as royalty. As already mentioned, neither yellow nor blue, and therefore also no green, are present in magenta. Magenta reigns supreme over all the colours. It is the intensification and indeed the spiritualization of all colours. It lends the colour circle a spiritual pole.

In terms of physics, magenta is an addition of vermilion and violet, the end colours of the spectrum. It overcomes its own darkness by means of darkness (the black band in the middle is replaced by magenta). Magenta comes into being by overcoming and integrating darkness.

In the rose of the alchemists magenta represents the quintessence of the four elements of earth, water, air and fire.

THE COLOUR CIRCLE

The colour circle is one accepted way of depicting all the colours in their totality. We shall here be introducing three colour circles: the one originating in the work of Johann Wolfgang Goethe; the one which Rudolf Steiner developed further on the basis of Goethe's work; and the one by the modern colour scientist Harald Küppers. We shall be referring only to those aspects of these colour circles which are relevant to the present work.

Goethe's colour circle

Goethe's most important contribution was to include magenta (which he called 'Purpur') in his circle. This colour not only provides the complement to green, but in its position at the top it also represents a culmination of all the colours. The hierarchical principle which characterizes the inner significance of colour is an important element in their organization. Colour circles in which magenta does not occupy the top position fail to recognize its special prominence.

It is interesting to note that in physics magenta is produced by the addition of vermilion and violet. And when a prism is used, magenta comes about by the lightening of the black band.

Green comes about by mixing yellow and blue in white. This causes subtraction, i.e. darkening.

Here we once again encounter Goethe's archetypal phenomenon: colours come into being by processes emerging from darkness or from light.

The concepts of addition and subtraction belong to modern colour science. Strictly speaking they are not Goethean because they reckon solely with light as the creative force. However, they do show that addition always moves towards white, the light, and subtraction towards black, the dark.

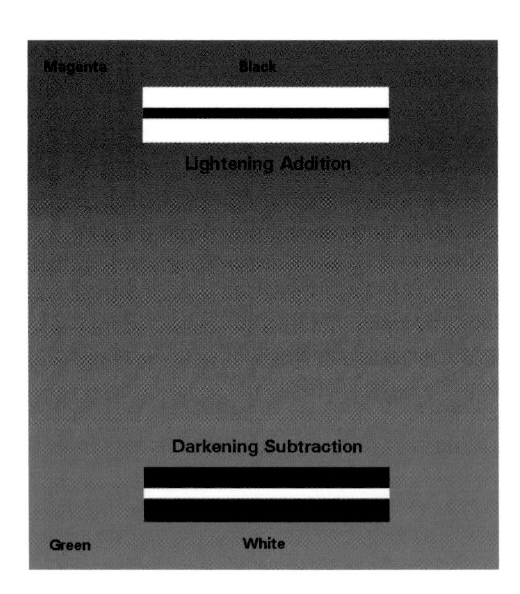

Magenta arises through the lightening of darkness: dark band seen through a prism. Green arises through the darkening of light: light band seen through a prism.

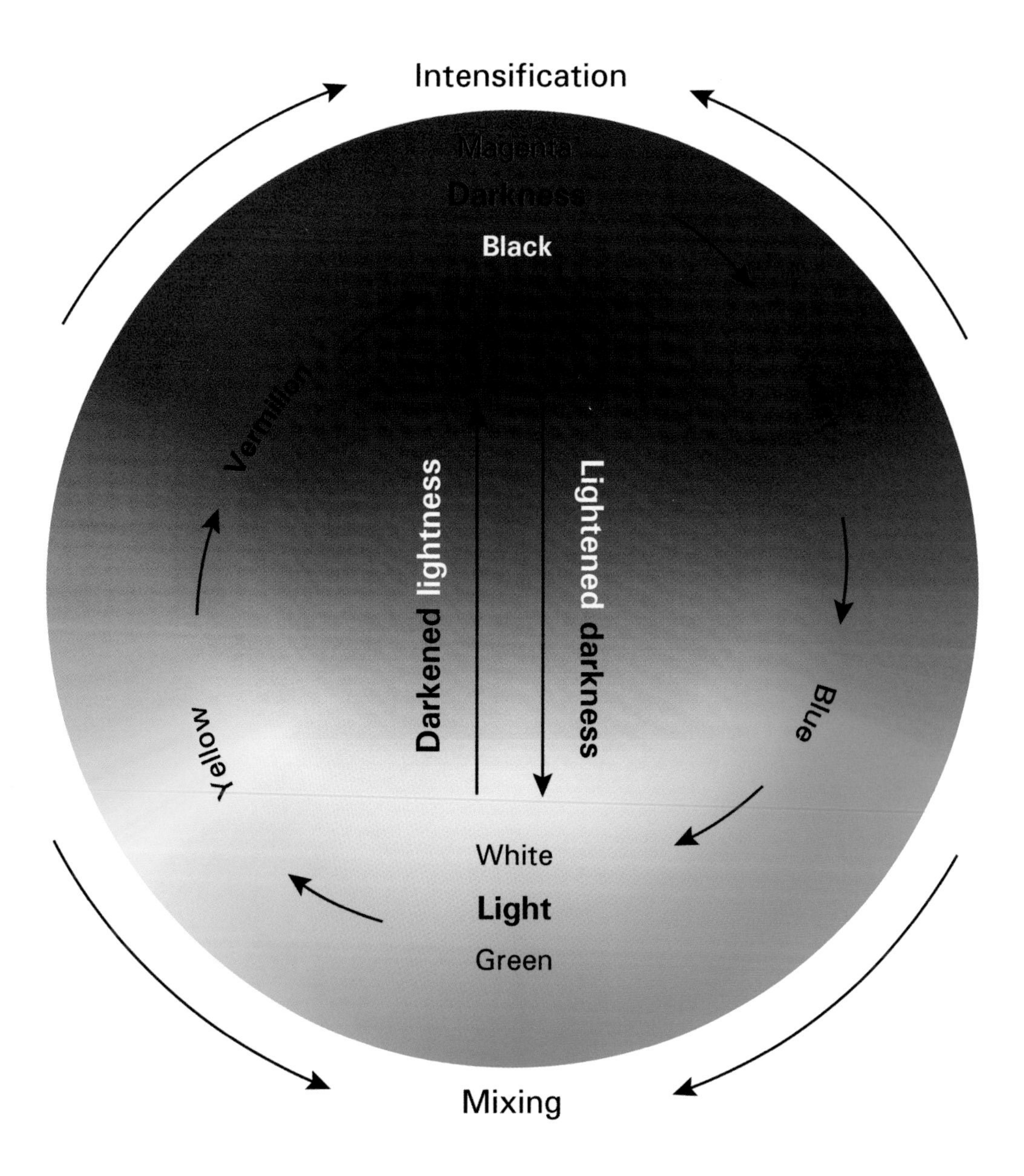

Goethe shows colours arising out of processes coming from light and from darkness and their encounter in mixing (green) and intensification (magenta).

Lustre colours and image colours

In Rudolf Steiner's terminology, colours arising through interaction from light or from darkness, i.e. yellow-vermilion and violet-blue, are called lustre colours. The creative power of light or of darkness still shines out of them.

And where colours have reached an end-point, as is the case with black and white, or a resting blend as in green, or a purified intensification as in magenta, Steiner spoke of image colours. They are a picture or an image of something.

The two groups of four colours are polar opposites both as phenomena and in the sense of spiritual science. Thus something arising through a process (lustre colours which are comparable to the planets) is complemented by solidified end-points (image colours which are analogous to the constellations of the fixed stars).

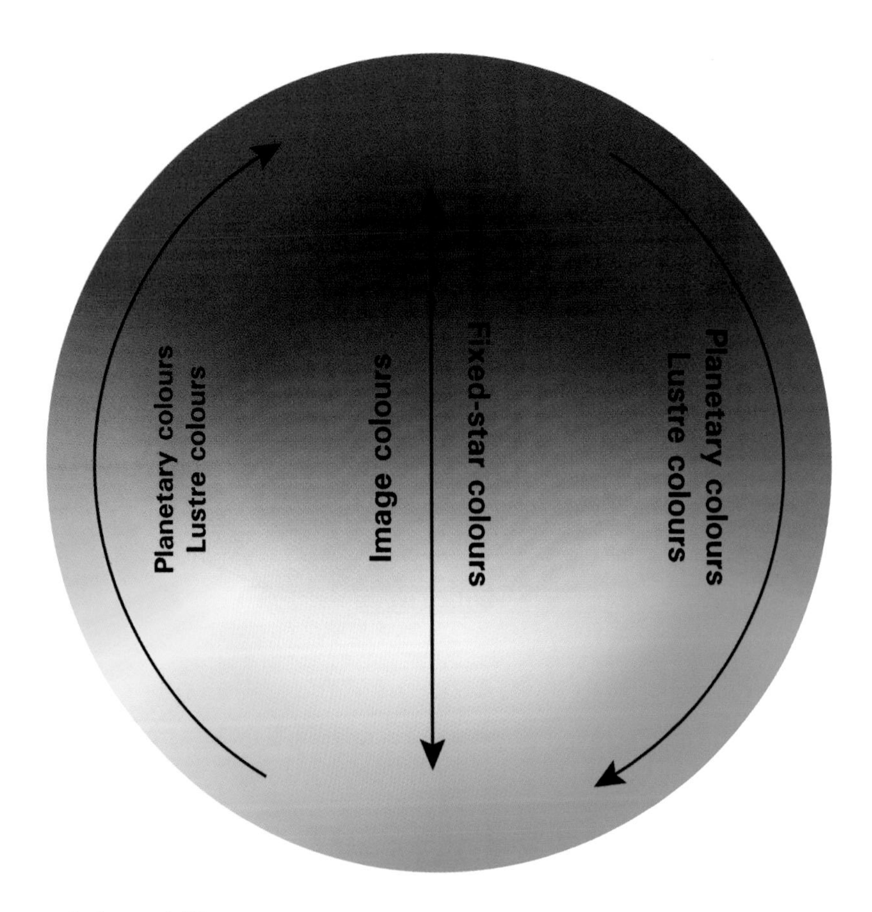

Colour circle by Rudolf Steiner.

Lustre colours and image colours according to Rudolf Steiner

Lustre colours	*Image colours*
Yellow and Red	Black and White
Violet and Blue	Magenta and Green
Sun	Moon
They shine forth.	They depict.
Through them shine light and darkness.	There is something shadowy about them.
In movement	Static
Planets	Fixed stars

Lustre colours

Yellow is the lustre of the spirit.
Red is the lustre of the living.
= shining lustre

Violet is the lustre of transformation.
Blue is the lustre of the soul.
= absorbent lustre

Image colours

(Light)	
White	The soul image of the spirit
Magenta	The living image of the soul realm
Green	The dead image of what lives
Black	The spiritual image of what is dead
(Darkness)	

Evolution

Old Saturn	*Old Sun*	*Old Moon*	*Earth*
Heat	Light		
	Lustre colours	Image colours	Earth colours

The colour circle according to Harald Küppers

Harald Küppers is the best-known colour scientist in Germany specializing in finding practical solutions for the technical handling of colours. His colour circle is not in opposition to that of Goethe. He also takes his departure from two times three colours together with black and white. Although he scorns Goethe's theory of colours he arrives at the same colour circle while explaining the colours on the basis of Newton's work.

Küppers derives the colours from the three primary colours vermilion, green and violet, which according to recent sensory research arise physiologically in the eye. In general it is obvious that today's colour scientists are increasingly interested in the physiology of colours, i.e. what the eye sees, rather than in theories of light or of colour. So there is little understanding of the living nature of colours on which a comprehension of the sensual perception of colours can be based.

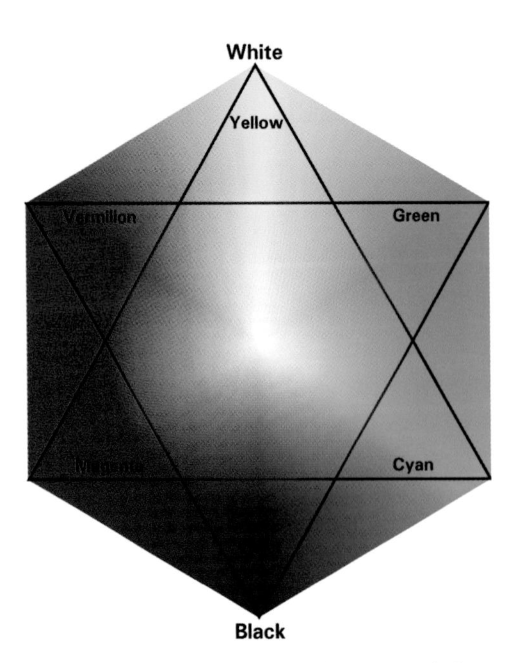

The lightest colours are at the top and the darkest at the bottom.

Küppers adds the three primary colours to the secondary colours magenta, yellow and cyan blue. Cyan blue is the scientific name for pure blue (without any magenta or yellow). Addition of all the colours yields white, and subtraction of all the colours yields black.

In this colour circle yellow is at the top because it is lightest and closest to white. Violet is the darkest colour.

Küppers transposes the eight colours on to a cube with one colour for each corner. Then he stands the cube on the black corner so that the white is at the top. He then distends the cube to form a rhomboid. This results in a coloured model which quantitatively depicts all the colours on the surface and the inside of the rhomboid.

It is important to distinguish between the two times three colours: the primary colours (violet, green, vermilion) and the secondary colours (magenta, yellow and cyan blue).

This model is indeed most useful, but it lacks the Goethean dimension of what the senses perceive.

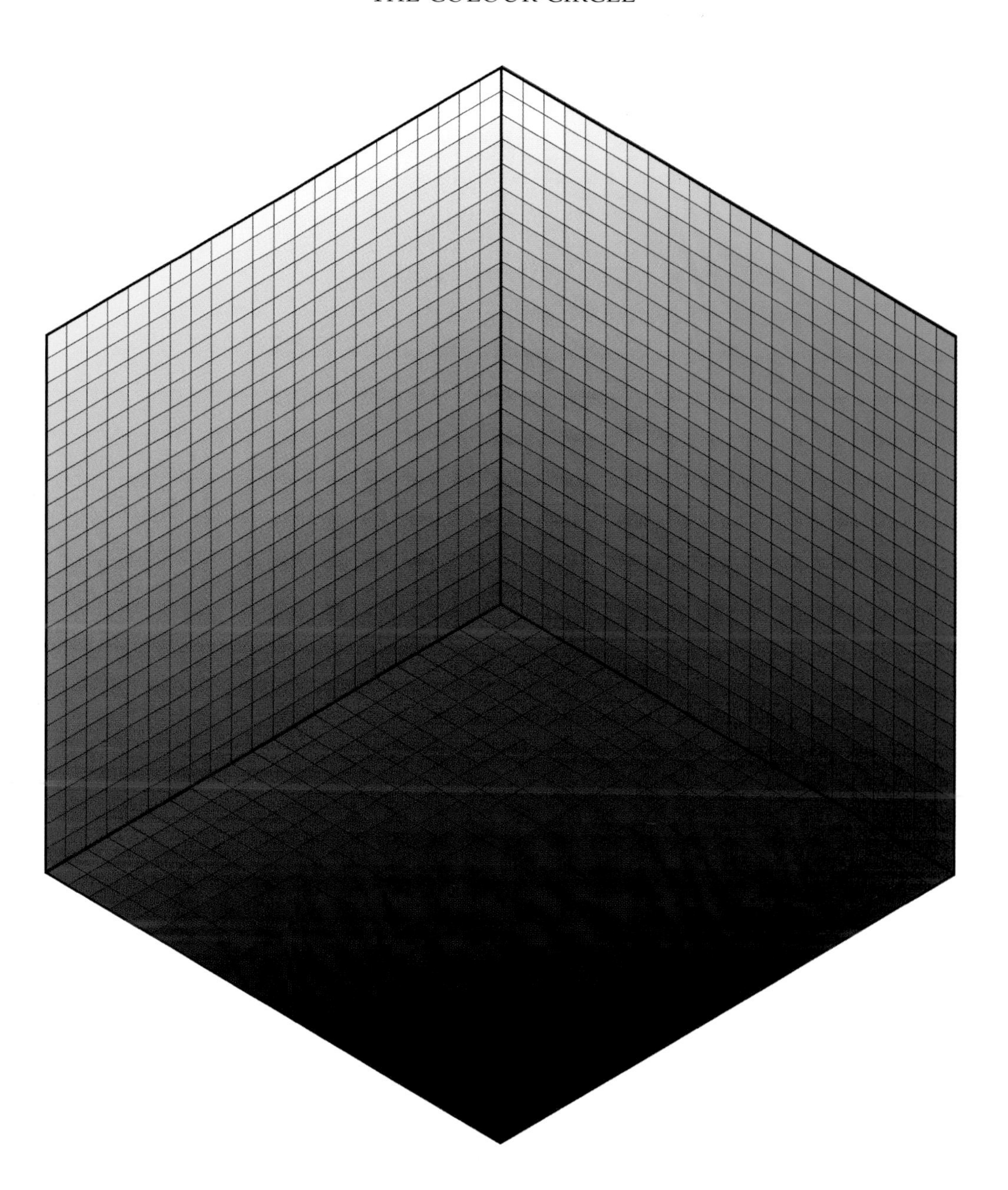

The Küppers rhomboid. All the colours, including white and black, are depicted on a distended cube, a rhomboid, which stands on one of its corners. All possible combinations, including grey and brown shades, are defined at specific geometric locations on the surface and inside the cube.

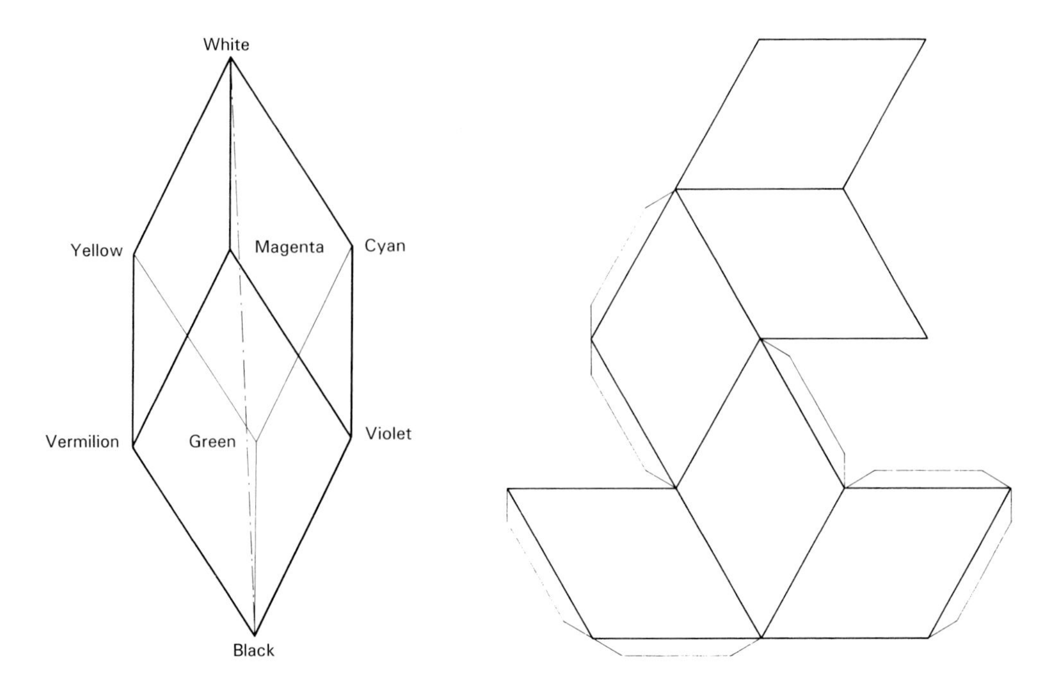

Rhomboid with vertical colourless axis (left). Exploded diagram of the rhomboid model (right). (From: Harald Küppers, *Das Gesetz der Farbe*, 1978.)

A hand-made painted rhomboid.

Caran d'Ache crayons

For non-experts it's useful to get hold of the correct colours when setting out to create a colour circle. The Prismalo Watercolour Crayons by Caran d'Ache are very practical since they are applied dry and can then be painted over with water afterwards.

For the following colour circle I have selected the 12 colours needed from among the dozens that are available. If you have the numbers and names of the colours they can be obtained singly from specialist art shops.

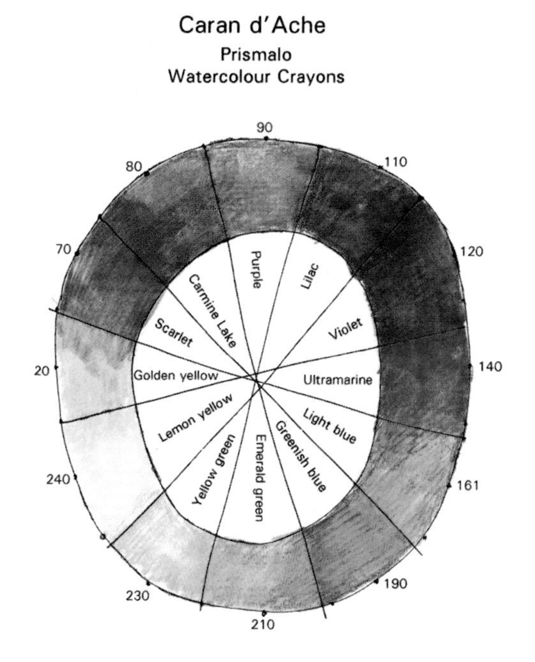

The Prismalo colours (numbered) selected here make for a beautiful colour circle.

Petals and leaves arranged in a colour circle.

A colour circle made of natural materials

Participants in my colour courses are often most impressed by colour circles using natural materials (petals, leaves, wood, stones). Collecting and arranging natural colours can become almost like a ritual, and the better the result the greater the fun assembling them will have been.

The nuances are infinitely variable depending on location and season.

THE THREE PRIMARY COLOURS

The three primary colours, vermilion, green and violet, are in a sense the archetypal colours. Even in his day Aristotle described them as such. They give structure to the rainbow with vermilion and violet forming the outer edges and green the middle. Where light is darkened and where darkness is lightened vermilion and violet border on the surrounding gloom.

Even in his day, Aristotle declared the primary colours vermilion, green and violet to be the archetypal colours. In the structure of the rainbow, vermilion and violet provide its borders while green appears at the centre. Where the light is darkened and where the darkness is lightened, vermilion and violet furnish the boundary. Beyond these boundaries we speak of infrared and ultraviolet. Where yellow and cyan meet, green forms the stillness in the middle realm. Research today states that the cells of the eye are specialized to see the colours vermilion, violet and green. And through addition to these, human beings can also perceive the other colours, yellow, cyan blue and magenta as well as all their combinations.

The three primary colours can also be separated out in a white band by means of a prism, if the refraction is strong enough to cause yellow and blue to disappear, leaving only violet, green and vermilion embedded in the black darkness.

Observed with the naked eye.

With greater refraction.

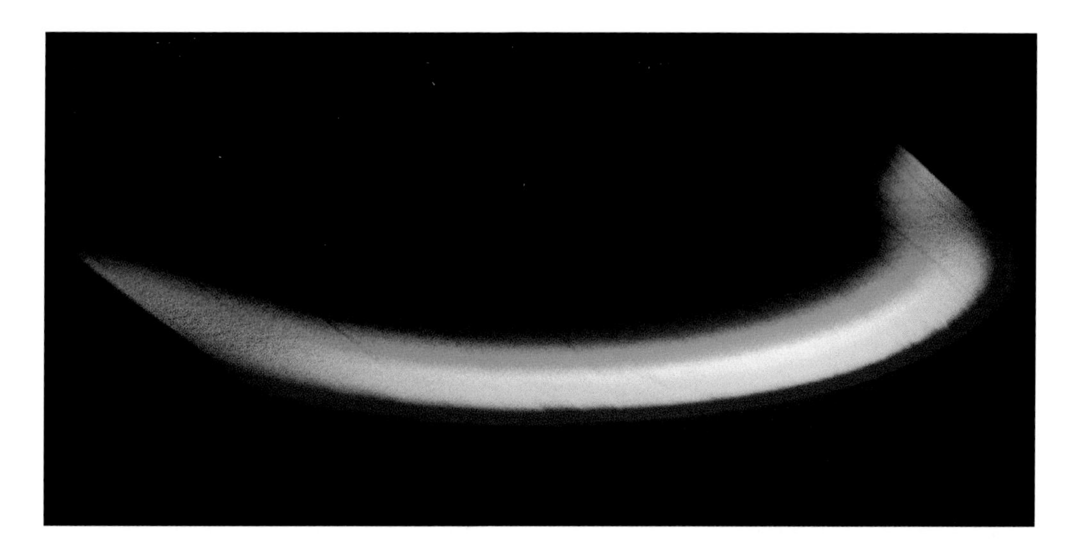

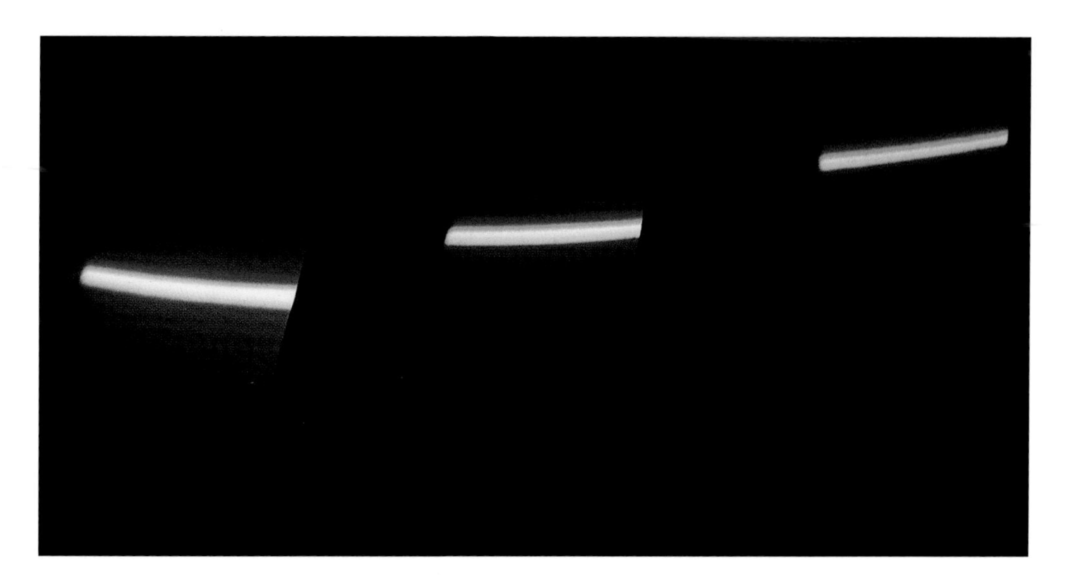

Objective experiments with the prism using a white band.

THE SECONDARY (OR PURE) COLOURS

Vermilion and violet together give magenta, vermilion and green give a pale yellow, and violet and green a pale cyan blue. All three together yield a colour that is almost white.

Magenta, yellow, cyan blue, white

The secondary colours, magenta, yellow and cyan blue, can be isolated in a black band by a prism if the refraction is sufficient to cause the violet and the vermilion to disappear, leaving only the secondary colours visible. The darkness of the black disappears and the secondary colours are framed by a bright white.

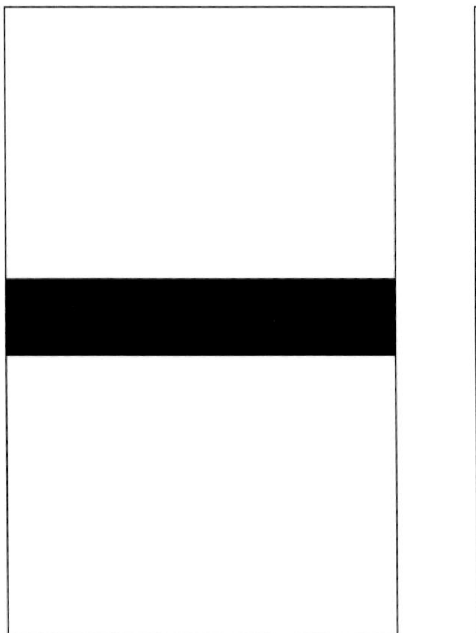

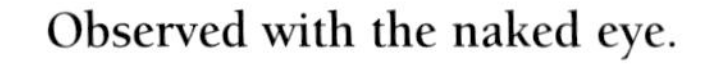

Observed with the naked eye.

Greater refraction.

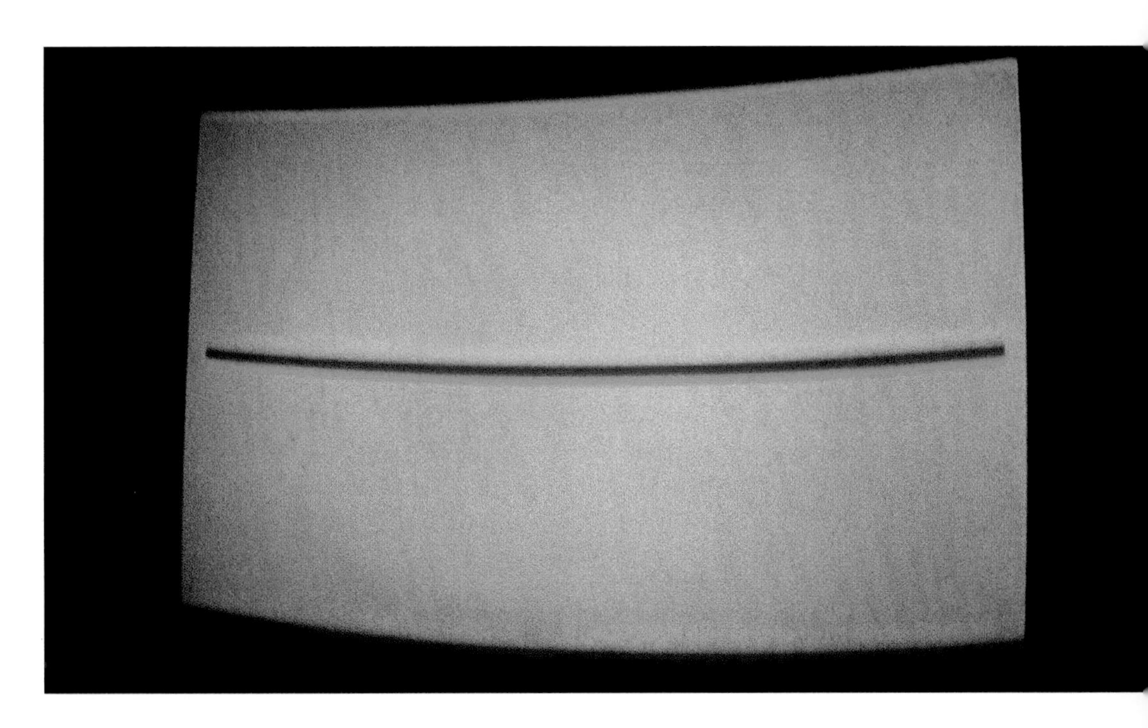

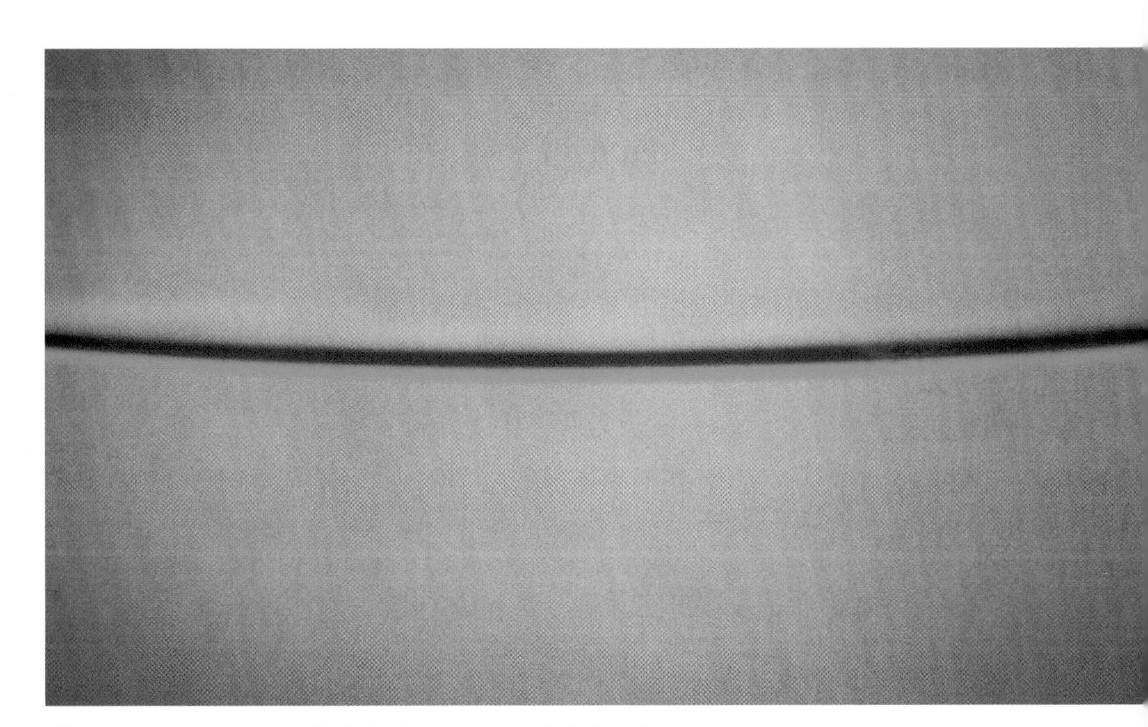

Objective experiment with the prism using a dark band.

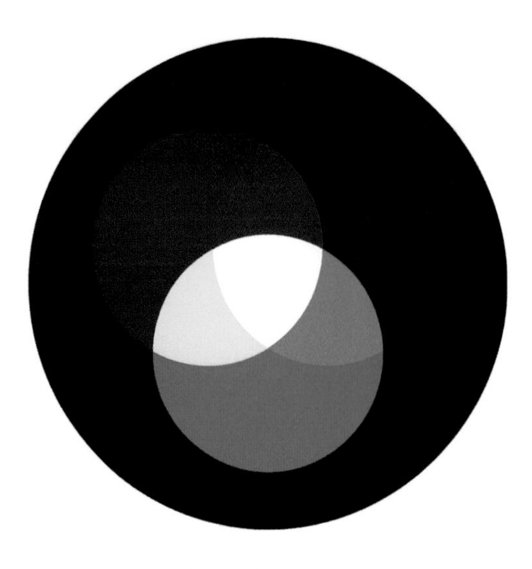

Vermilion, violet and green projectors reveal the secondary colours by addition.

Creating magenta, yellow, cyan blue and white by means of addition

Magenta, yellow, cyan blue and white can easily be 'produced' optically by addition of the primary colours violet, vermilion and green, when each has its own light source.

Vermilion and violet yield the secondary colour magenta. Vermilion and green yield a pale yellow. Violet and green yield a pale cyan blue. All three together are almost white.

When you project the three colours (using three projectors) so that the coloured surfaces overlap, the secondary colours are the result (see illustration).

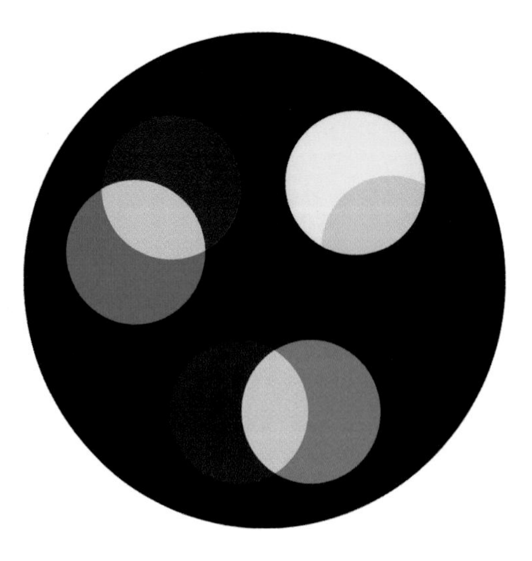

Two colours each with its own light source are projected one on top of the other. This addition neutralizes the colours.

Neutralization by addition of complementary colours in pairs

Those inexperienced in working with colour will be surprised by the way in which the secondary colours come about through addition of the primary colours vermilion, green and violet. We have to digress in order to explain this. If we combine the opposite pairs of the colour circle by addition, the colours disappear or are neutralized. They become colourless, grey (uncoloured).

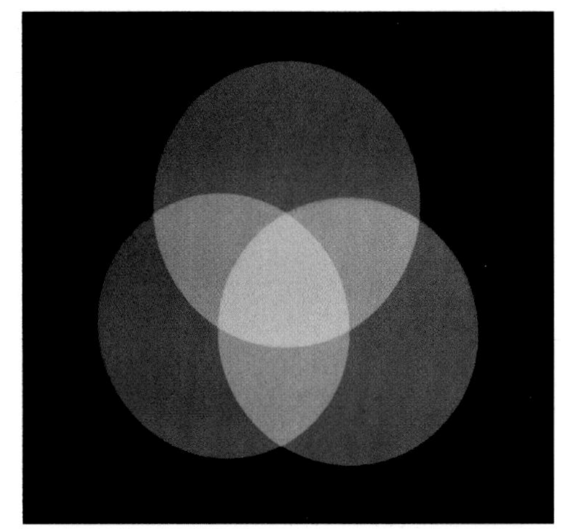

Various additions.

How the secondary colours come about through addition

Any two colours which are complementary to one another disappear into colourlessness when mixed. All three primary colours added together give colourless light, i.e. white.

These colour equations show how the secondary colours arise from the combined colours by addition of quantities of light. The colours which neutralize one another, which are always opposite one another in the colour circle, yield colourless, i.e. relatively white light. The colours which remain are the pure light colours. The more light qualities included, the brighter or whiter does the white become, i.e. a very pale grey.

Magenta, yellow, cyan and white are, one might say, the basic colours from which all other colours arise through addition or subtraction. There is always some darkness (grey) in white. The whitest white can become even more white through the addition of even more light quantities.

Vermilion	+ Violet			= Magenta
Vermilion	+ Cyan blue	+ Magenta		= Magenta
Neutralization	+ Magenta		Magenta	= Magenta
Vermilion	+ Violet			= Magenta
Magenta	+ Yellow	+ Violet		= Magenta
Magenta	+ Neutralization			= Magenta
				= Magenta
Vermilion	+ Green			= Yellow
Vermilion	+ Cyan blue	+ Yellow		= Yellow
Neutralization	+ Yellow			= Yellow
			Yellow	= Yellow
Vermilion	+ Green			= Yellow
Yellow	+ Magenta	+ Green		= Yellow
Yellow	+ Neutralization			= Yellow
			Yellow	= Yellow
Violet	+ Green			= Cyan blue
Violet	+ Yellow	+ Cyan blue		= Cyan blue
Neutralization	+ Cyan blue			= Cyan blue
			Cyan blue	= Cyan blue
Violet	+ Green			= Cyan blue
Cyan blue	+ Magenta	+ Green		= Cyan blue
Cyan blue	+ Neutralization			= Cyan blue
			Cyan blue	= Cyan blue
Vermilion	+ Violet	+ Green		= White
Magenta	+ Yellow	+ Violet	+ Green	= White
Magenta	+ Green	+ Yellow	+ Violet	= White
Neutralization	+ Neutralization			= White
			White	= White

THE MIXED COLOURS

The mixed colours arise by means of subtraction of the secondary colours magenta, yellow and cyan blue, and also white. This is especially applied in printing. The three secondary colours at least are needed in order to print a coloured picture. The white colour of the paper and the black of the printer's ink provide structure for the picture.

The term colour subtraction is applied when there is only one light source, e.g. a single projector (white light) or the white paper on which the print is to be imposed. The secondary colours are superimposed on one another by the use of coloured filters so that the quantity of light is reduced. The colours themselves are added as usual. The more the number of coloured filters that are superimposed on one another, the darker they become. If they are neutralized by complementary colours, the resulting tints are grey to black.

In Goethean terms, subtraction denotes the removal of light quantities. It could also be described in positive terms as the addition of darkness quantities.

Black is the same thing as a dark white. The more the quantity of light passing through the filters is reduced, the darker does the white become. It is then described as pale, medium or deep grey or, indeed, black.

Shades of brown

Various shades of brown come about through the subtraction of vermilion and violet, green and vermilion, and green and violet.

The subtractive mixtures become darker or lighter depending on the amount of white and black that is added.

Yellow	+ Magenta	+ Cyan blue	= Black
Vermilion	+ Cyan blue		= Black
Neutralization			= Black
		Black	= Black

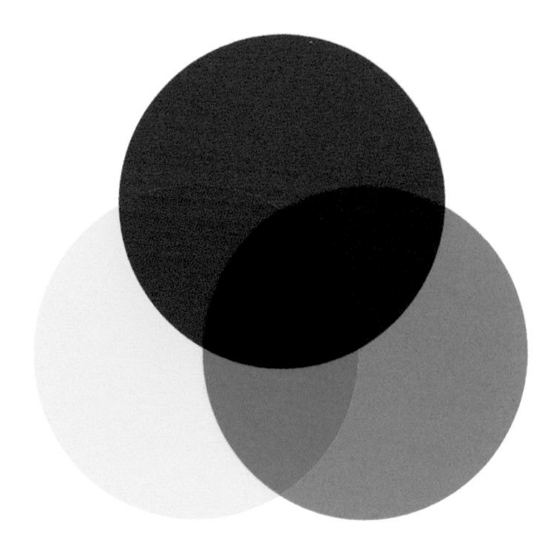

Mixtures (subtraction) of secondary colours yield the primary colours of the spectrum.

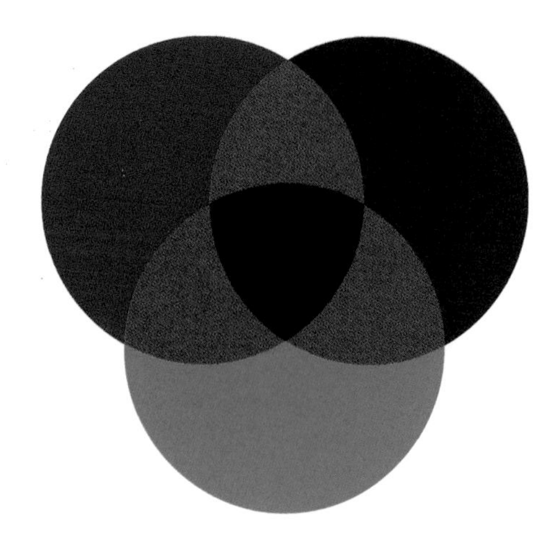

Mixtures (subtraction) of primary colours yield shades of brown).

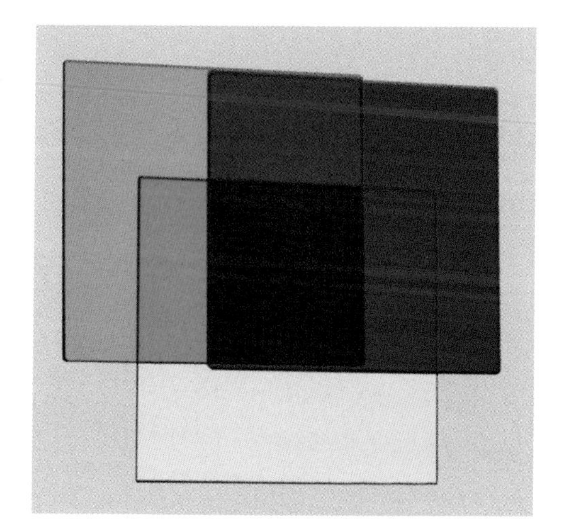

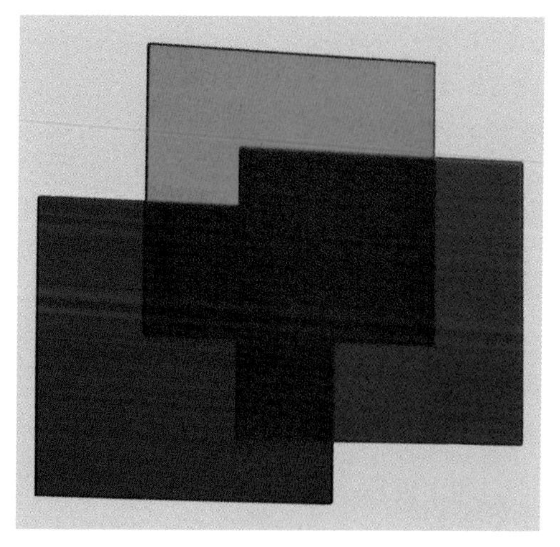

Subtractions yield shades of brown.

Cyan, magenta, yellow and black are superimposed on one another in four-colour printing. The individual colour proofs are shown on the right.

DELAYED AND SIMULTANEOUS CONTRASTS

Delayed and simultaneous contrasts take us into the shadowy realm of the eye socket, to the colours seen as a flickering on the retina when there is absolute darkness. When we look momentarily at a bright light, for example the sun, and then close our eyes, we see colours which are projected on to the inside of our eyelids (or on to the world around us if our eyes remain open). Stimulated by excessively bright light the retina reacts to the overly strong effect. In fact our eyes endeavour to heal or neutralize the excessive stimulation by reacting with the appropriate colours. Remarkable agate-like rings of colour arise which are constantly in flux, often in a complementary way, and gradually decrease in size.

Our eyes dislike one-sidedness, so they constantly seek a totality. When they receive one-sided impressions of colour or light they react with healing complementary colours that restore the totality.

But we must distinguish between delayed contrasts which arise one after the other, and simultaneous contrasts which appear instantaneously.

Delayed contrasts

Colour contrasts arising as a reaction to bright light are a typical reaction of the retina seeking to adjust to or heal the impact.

When we make demands on our eyes not by a bright light but by the continuous impression of a one-sided colour, the retina reacts with the complementary colour.

Look for a minute at one of the black dots within a coloured field on page 86 and then look to the right at the black dot in the white field.

You will see the complementary colour, only much brighter and more ethereal. You can also project this complementary colour on to the palm of your hand or on to a white wall. The further away the surface, the larger will the after-image be. As long as the background is colourless, the complementary colour will arise, i.e. the one that is situated on the opposite side of the colour circle (see page 85).

If you project the complementary colour of the after-image on to a coloured background, a corresponding mixing will occur.

The appearance of the ethereal complementary after-image colour is delayed. It remains until the retina has had time to regenerate itself.

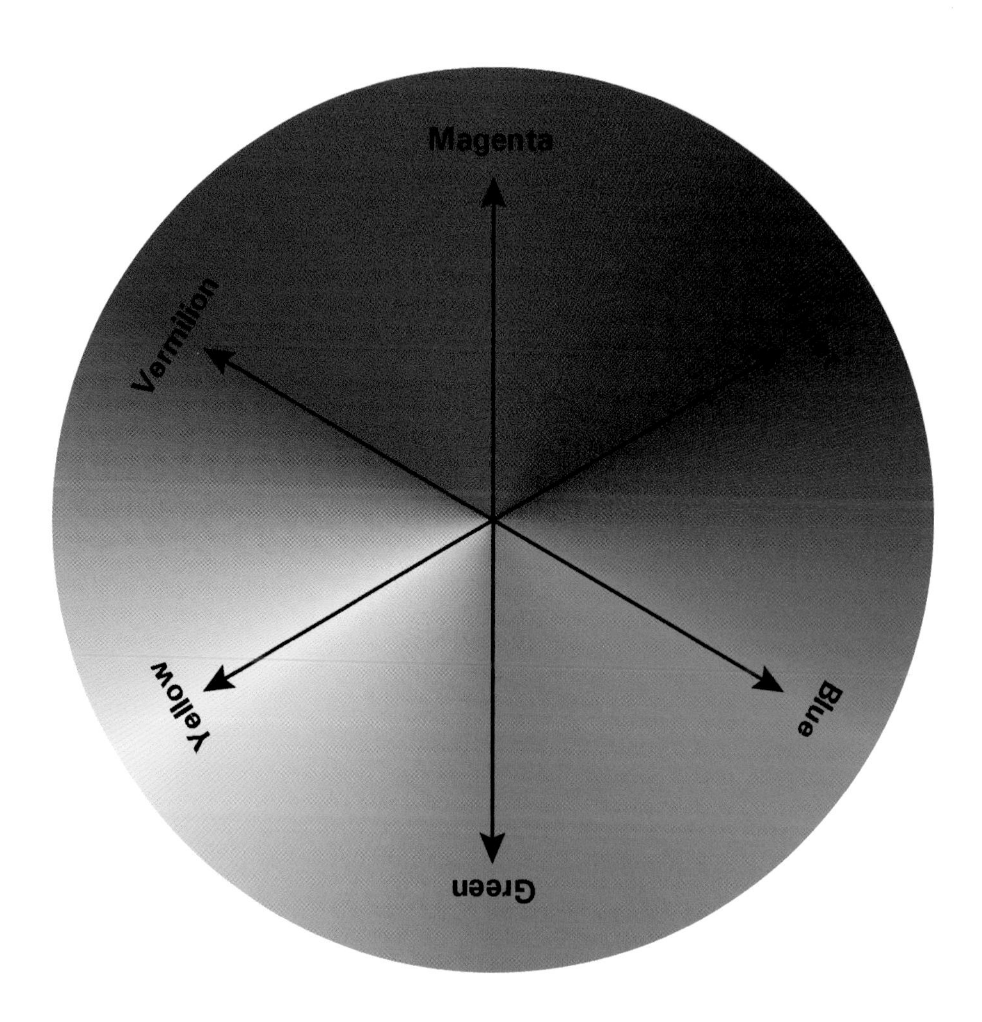

The colours opposite one another in the colour circle are termed complementary colours.

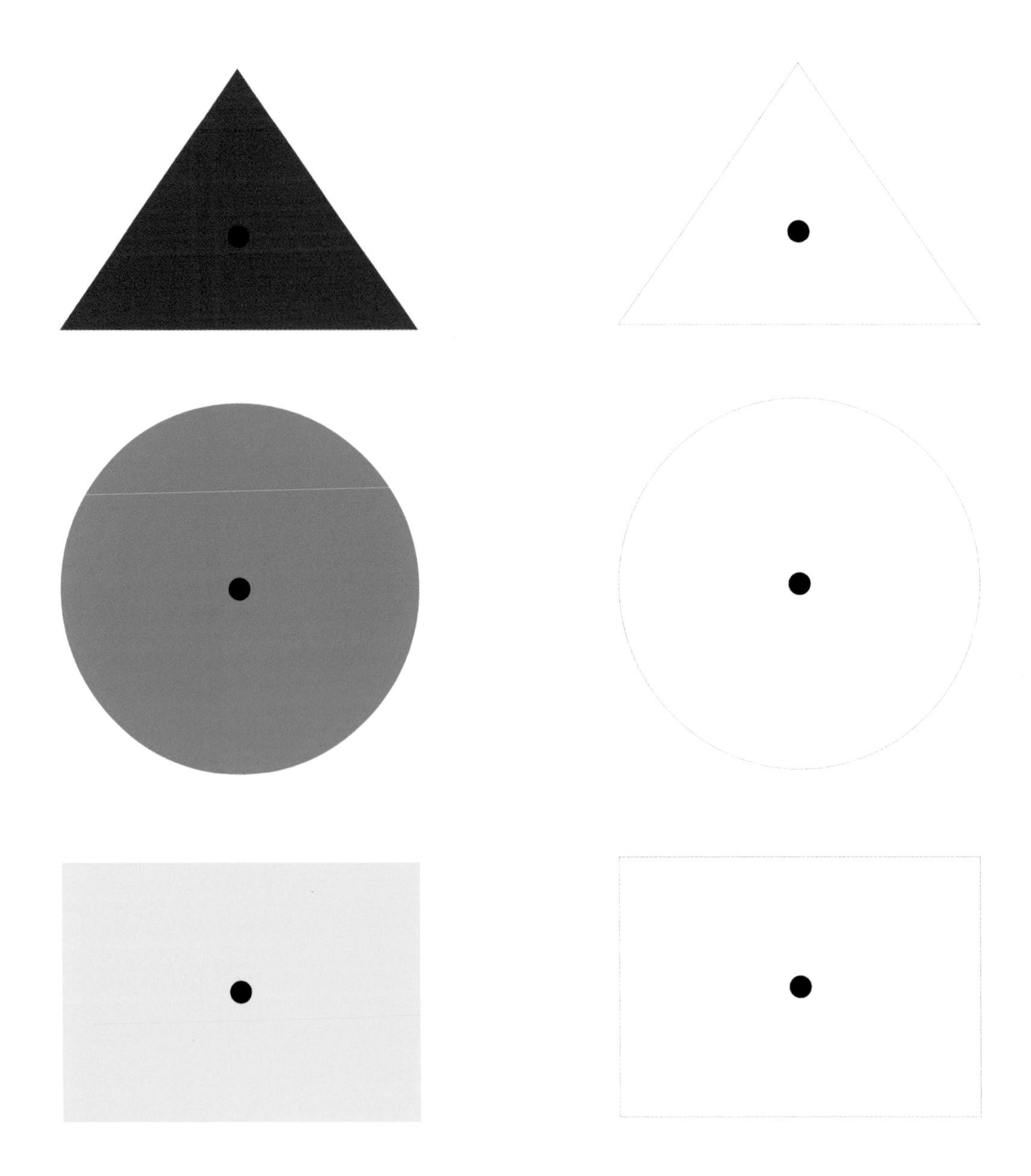

Look for a minute at the black dot in one of the coloured fields, then at the adjacent dot.

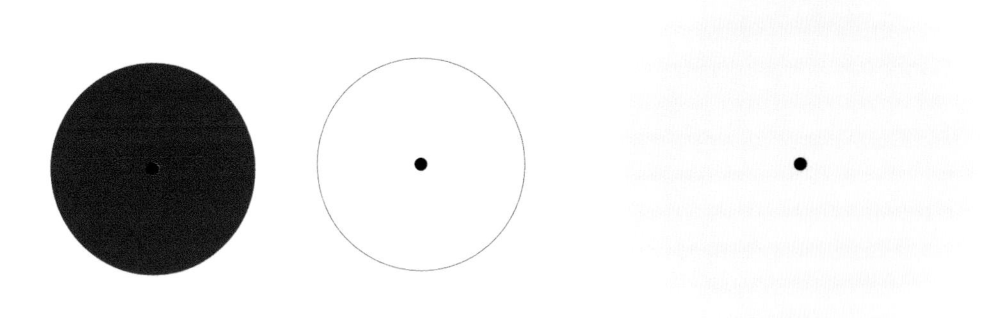

Look for a minute at the black dot in the red field, then at the dot in the white field, and then at the dot in the yellow field.

The appearance of the ethereal complementary after-image colour is somewhat retarded, so it is termed a delayed colour. The retina takes time to regenerate itself and then the colour fades.

Complementary scales

Light scale

Black	Violet	Blue	Green	Red	Orange	Yellow	White
0	1/4	1/3	1/2	1/2	2/3	3/4	1

Heat scale

Warm colours							Cool colours
Warm	Orange	Yellow	Red	Green	Violet	Blue	Cool
1	6/7	5/7	4/7	3/7	2/7	1/7	0

A coloured projector generates black shadows. When these shadows are lightened, they become coloured, taking the colour which is complementary to that of the projector.

Simultaneous contrasts

The phenomenon of the simultaneous contrast which appears instantaneously as a supplementary colour is astonishing in the world of colour and unexplained by colour science. Good examples can be observed in coloured shadows.

When you project a red light on to a white wall in a dark room, the wall becomes red. When an object casts a shadow on the wall, the shadow is black. When you lighten this black shadow with a white light from another projector which is not too bright, the shadow immediately turns green. One could say that in the shadow surrounded by red there slumbers the complementary green which becomes visible when the darkness is lightened. This exact complementary colour of the shadow can be generated with every colour. And the coloured shadow can also be photographed. It appears and disappears together with the surrounding projected colour. It is perfectly clear and objective and not at all ethereal or brightened like the delayed contrasts on the retina.

Scientists disagree as to the reason for the colour of the shadow. Is it generated by the retina as Goethe suggested, since he counted the coloured shadows (green snow at sunset) among the physiological colours, i.e. the colours in the retina. Although normally not in agreement with Goethe, in this case science does agree with him. Rudolf Steiner, on the other hand, disagreed with Goethe and expressed the opinion that the colour is objective, i.e. independent of the eye.* (When, by the way, Eckermann, Goethe's secretary, made the same objection, the great man reacted indignantly.) Rudolf Steiner's opinion has also been questioned, for it is a fact that when you view the shadow through a tube, thus isolating it from its coloured surrounding, it immediately loses its colour.

The colour of the simultaneous contrast always appears instantaneously as the complementary colour of its neighbour: a grey beside red appears greenish, and beside green reddish. The eye immediately endeavours to make a totality out of the colour and lightness combinations. Is it the eye (retina, brain etc.) which brings about the totality, or is it the colour or the lightness or darkness? Whatever the case may be, the fact is that the world of colour always endeavours to show itself in a totality. That is why colour has the property of creating a totality – something of which we are nowadays so very much in need.

---

* See further in Rudolf Steiner's lecture of 30 December 1919 in *The Light Course* (Anthroposophic Press 2001); also note 1 to the same lecture.

Coloured shadows.

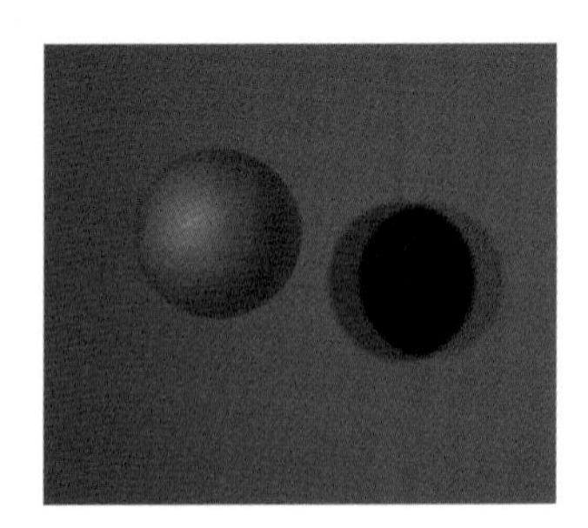

Shadows thrown by coloured projectors are lightened by candlelight.

Vincent van Gogh, a magician of colours in light and darkness

In the context of the complementary colour contrasts we must surely cast a glance at the life and work of Vincent van Gogh. More than in the case of any other painter, his theoretical and above all his painterly remarks about colour polarities lead us directly into mysteries of the colour world. As we know from his life, the polarities and tensions were also an existential experience for him. In combination with the darkness of earthly suffering his sunny and fiery spirit made a true communion of heaven and earth.

During his brief life he met with the Impressionists, those painters of light. He was fascinated by the unsullied colours of their sunshine, the brightness and joyousness of their landscapes. He experienced how it was possible for a painter to come close to all that is ethereal and colourful in nature. He became familiar with the colour contrasts of the complementary colours and began to doubt whether it was permissible to include black in his palette of paints. As a preacher in the mines of the Borinage he had been all too familiar with the existential sufferings of others, as well as his own, and had applied earthy blacks and browns in *The Potato Eaters*, one of his earlier works. But now he experienced the painting of brightness and light as a

Vincent van Gogh, *Potato Eaters* 1885 (Rijksmuseum, Amsterdam).

Vincent van Gogh, *Starry Night*, 1889 (Museum of Modern Art, New York).

release. He travelled southwards to Arles and St-Rémy in order to enter into the clarity of light and the blaze of the sun not only through his outward senses but also through his inward sensibilities.

His initial thoughts about a theory of colours were prompted by the painter of the Romantics, Eugène Delacroix, whose colour triangle with the three main colours, red, yellow and blue, and the intermediate colours, orange, green and violet, corresponded exactly to

Vincent van Gogh, *Vincent's Bedroom at Arles*, 1888 (Rijksmuseum, Amsterdam).

Goethe's way of thinking, except that where red was concerned he did not explicitly mention magenta or purple.

At Arles van Gogh distanced himself from the Impressionists and permitted white and black to re-enter his range of colours. He considered that as the end-points of the colour sequence, white and black represented lightness and darkness. In this way black and white became points of rest for the actual colours as they moved.

For the colours do move, for example, on the complementary scales of orange/blue, green/red, and violet/yellow. When they are at opposite ends they intensify one another. But if they mingle they dim their strength or even extinguish one another – as do fire and ash – in a middling grey.

Black also serves to separate objects from one another, something the Impressionists did not like, since nature lacks demarcation lines of this kind. Van Gogh, however, did use black contours to separate objects from their background as, for example, in the picture of his bedroom, thus giving

Vincent van Gogh, *Sower at Sunset*, 1888 (Rijksmuseum, Amsterdam).

each object its individual, detached existence. In this way he individualized the subjects he painted, also by his use of light and darkness and complementary colours. High and low tints, pale and deep colours (*L'Arlésienne*); solitary destiny yet bound up with the cosmos as a whole (*Portrait of a Poet*); the starlit heavens and the earthly plane (*Starry Night*); and a painter who consumes himself like a phoenix before time and again rising refreshed and more beautiful than ever from the ashes (*Self Portrait*, September 1889). Thus does a painter's destiny become a living example of a colour theory.

In *Sower at Sunset*, van Gogh painted both the sun and the earth, with the sower and the tree acting as mediators between the two worlds. To bring the sun down into the earth (sower) and raise the earthly realm up to the sun (tree) is an archetypal Christian motif. But the painting also expresses the suffering of the (as yet) unredeemed earth. The light shines into the darkness, but the darkness has (as yet) failed to comprehend it. Vincent van Gogh places his faith in the sun and the earth. For him sun and earth are the centre of creation.

Even though since Copernicus the sun has taken earth's place as the centre of creation, Vincent van Gogh also accepts the dark earth of destiny without, however, negating the sun. The great astronomer Tycho Brahe (1546–1601) much earlier depicted the integration of the two centres in his polar image of the world. Here the earth is the centre of creation while sun (and moon) circle round it. But the planets circle round the sun. This is more than just a mental game, for it is the destiny of mankind not to become totally detached from the earth.

We admire van Gogh's paintings because they picture the individual realms. They contain both sun and earth, light and darkness, enthusiasm and the gravity of destiny. Although he broke away from the Impressionists, he did not separate himself off from nature. To him nature meant more than mere colourfulness; it was a meaningful symbol. But he was not comfortable with abstract symbolism, which in part was anyway no more than superficial decoration. He needed truly living destinies of human beings, of trees, landscapes and the skies, in order to join forces with them. The real meaning of incarnation is to unite fully with the earth while not allowing one's spiritual fire to be extinguished by it. Vincent van Gogh spiritualized the earth with his glowing colours. This, surely, is a labour of Christian redemption.

In his inimitable way van Gogh also characterized seasonal colourings with his use of complementary contrasts.

Here we see once again the eight basic colours, always as complementary pairs, as they relate to the seasons.

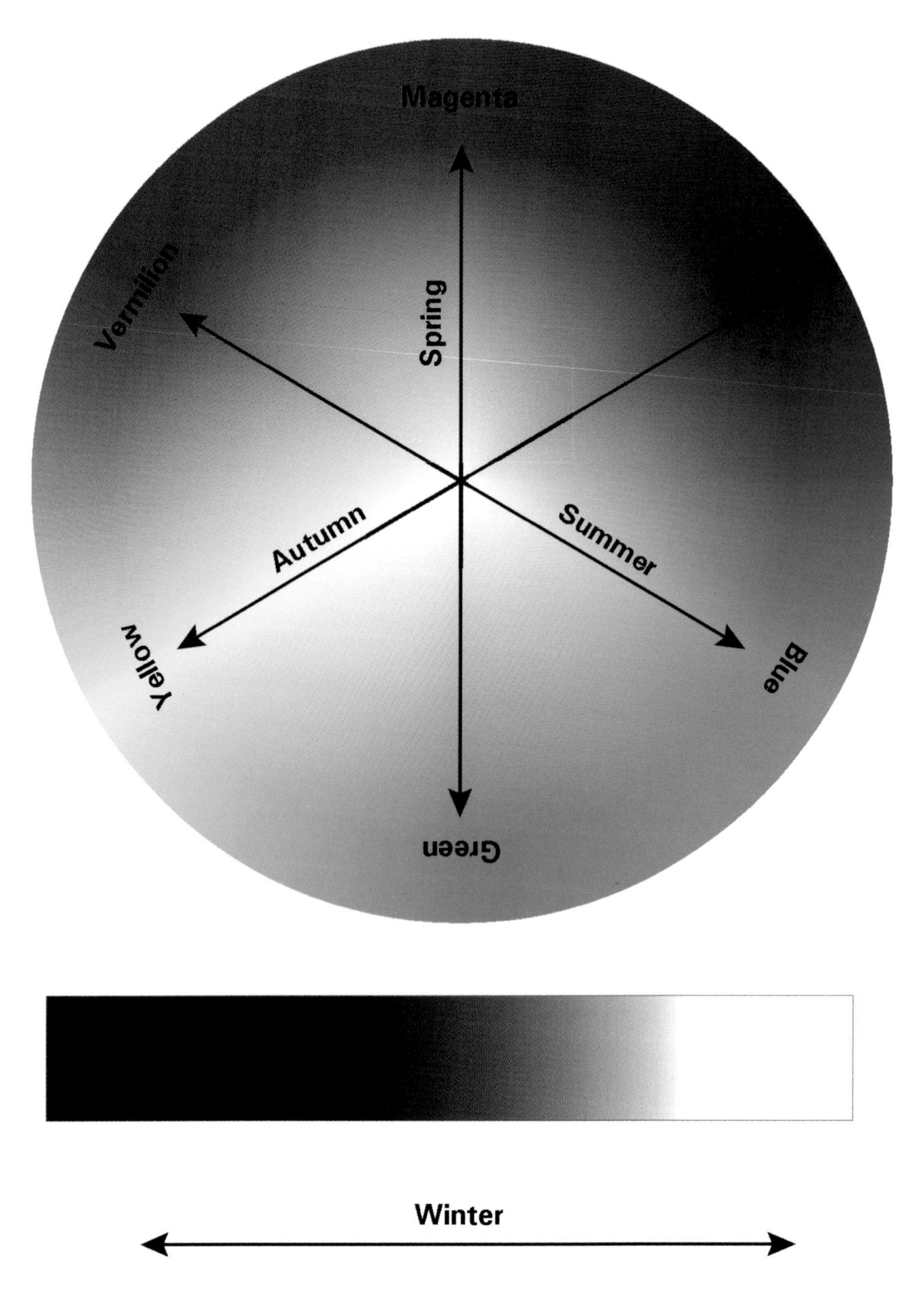

Magenta and green as signposts to a new ethic

In concerning ourselves with colours we can also be led into ethical realms. Good and evil, seen as the light and the dark, immediately reveal the ethical dimension.

Isaac Newton saw light as the origin of all colours. For him, darkness was simply the 'not light', just as evil was the 'not good' for Augustine, one of the fathers of the church. But if evil, that which is dark, is simply denied, if it ought not to be here and yet keeps on showing its face, the result is an ethic involving the repression of darkness. Erich Neumann spoke of this as being an 'outdated ethic'. It creates a split between good and evil, between spirit and physical urges, between church-going Christianity on Sunday and everyday egoism. It creates scapegoats that need to be destroyed.

Instead, Neumann formulated a 'new ethic' which integrates good and evil, light and dark, just as does the black-and-white magpie in Wolfram von Eschenbach's *Parzival*. The new ethic requires that, rather than being good or bad, people should be autonomous in respect of both good and evil. Evil that remains in the unconscious causes badness, whereas when it is brought into consciousness it brings about good. We need to experience evil if we are to become mature, as is described in Goethe's *Faust*, which integrates the good (Margarete) and the evil (Mephistopheles). Those who persevere in endeavouring to integrate the evil are the ones who will be saved! The new ethic requires of us that we should have the moral courage to be not only not worse but also not better than we actually are. It is good to recognize one's own evil, but to make oneself out to be better is bad. Neither submission nor self-importance can lead to a living wholeness within us. Such wholeness lives in the tension between good and evil, male

Jan van Eyck, *The Marriage of Giovanni Arnolfini and Giovanna Cenami*, 1434 (National Gallery, London). Here green and magenta unite with one another.

and female, black and white. This ethic of integration was described by Heinrich Pestalozzi in his work on the course of nature in the development of the human race. He showed that there is no need for individuals constantly to deny their nature (desires) in order to fit in with social norms by constant moral (ethical) effort since it is better to integrate the two. The moral force that is found within the individual provides the desires and the social norms with a holistic formative power. One hundred years later, Rudolf Steiner arrived at the same conclusion in his book *The Philosophy of Freedom*. To describe the moral force he used the terms ethical individualism, moral imagination, or simply intuition.

So what is the connection between the colours magenta and green? Goethe's theory of colours is holistic in a way that Newton's is not. For Goethe the forces that create the colours lie in light and dark. Violet and blue come from the dark realm while yellow and vermilion come from the light realm. But the decisive factor lies in the green which comes about when yellow and blue are mixed. In this way green integrates the colours of lightness and darkness. As we have seen, this occurs in the narrow white band by means of a prism, or in the middle of the rainbow, or in the plant that stands between the cosmos and the centre of the earth. Thus green becomes an image of life (R. Steiner); it is the colour that spreads nature across the earth and that brings individualization into the total spectrum ranging from vermilion (desires) to violet (asceticism). It brings the end-points of the spectrum into the resting middle point, just as is also obvious in the green colour of the heart chakra. Green becomes the integral colour of Christ because it unites heaven and earth.

Magenta rises above the colours of the spectrum. Seen through a prism it comes about as a lightening of the black band. Here vermilion (desires) and violet-blue (asceticism) are intensified to become magenta. In magenta darkness is lightened (addition), the spirit shines into the darkness, the soul blossoms when the human being blushes. Thus magenta becomes the image of the soul (R. Steiner), which can be intensified through intuition, through moral imagination to become ethical individualism. Magenta points towards the process of incarnation and of the integrated higher self. A new ethic reckons with individualizing colour processes such as this which intensify into a higher self.

As a pair of opposite colours, magenta and green appear in soap bubbles, in the sheen of spilt petrol in puddles, at sunset in a snowy landscape where the shadows are green, in the magenta of the dawn sky with pale green along the horizon. As the Russian Symbolists Vladimir Soloviev and Andrei Belyi put it, aurora, the red of the dawn sky, gives a magenta image of new beginnings and departures. The spiritual colour of magenta needs the vitality of green in order to become effective in the world as a whole. While magenta has to arise anew all the time above the top of the colour circle, like the power of morality, so green individualizes light and dark to become the secure pole of the heart.

COLOUR ENERGIES, VISUALIZATION, CHAKRAS

Colour meditations rest on the ancient tradition that holds colours to be active in our energetic, psychological and spiritual consciousness. The seven coloured energy centres of the human body, known as chakras, represent a totality of the world of colour.

The colours and the chakras

Colours are at work in our energetic, psychological and spiritual consciousness. It is on this ancient knowledge that the colour meditations are based. The seven coloured energy centres of the human being, known as chakras, present a totality of the world of colours. Each chakra receives the relevant energy, transforms it and then passes it on. So we have the colours of the rainbow within ourselves, the lowest being the root chakra with vermilion. It mediates physical energy and life forces, and it also grounds us. It gives us strength but also rage, love and hate. It rouses us to glowing emotions and strengthens our courage for optimistic deeds. When raised to consciousness it can work against depression, lethargy and lack of energy.

The next above this is the sacral chakra with orange as its colour. Orange is a mixture of red and yellow, so it creates balance between sexual arousal and the purely intellectual. It harmonizes restless and excessively nervous individuals. Orange stands for vitality and spiritual energy. It mediates a warm, cheerful and joyous energy and promotes the ability to make relationships and also to move about freely.

The next is the navel chakra, also known as the solar plexus, which is yellow. It stands for light, life and immortality. Yellow is actually the colour of the spirit because it has the greatest proportion of light. It speaks to those who possess creative gifts. A sense of happiness rays forth from yellow, so it is helpful for those who have a melancholic and pessimistic disposition.

The heart chakra is green. This colour occupies the middle of the rainbow and lies between the warm and the cool colours. It mediates between the material and the spiritual aspects of the human condition. It promotes growth but also stillness and equanimity. It vitalizes the blood, regulates blood pressure and heals heart problems. It also relieves headaches.

The throat chakra corresponds to the colour blue, the soul colour of depth, introversion, melancholy, fidelity and devotion. The heights of heaven and the depths of the ocean are the locations to which meditative absorption can take us. Blue brings about boundless peace and rest within oneself.

The deepest colour, indigo, rays forth from the third-eye chakra. It helps to purify the mental and psychological powers of the human being. It helps to connect us to the dark depths of the unconscious. Indigo is the midnight colour, known to alchemists as the 'black sun'. Since ancient times the deepest and highest meditative exercise involved seeing the light through the darkness.

Violet, which lies close to and yet far from vermilion, mediates the crown chakra. Violet has the shortest wavelength. As a mixed colour it mediates knowledge, passion and 'mystical facts'. It mediates the capacity to recognize open secrets in the midst of everyday concerns or, as Joseph Beuys put it, to recognize 'mysteries in the midst of the railway station'.

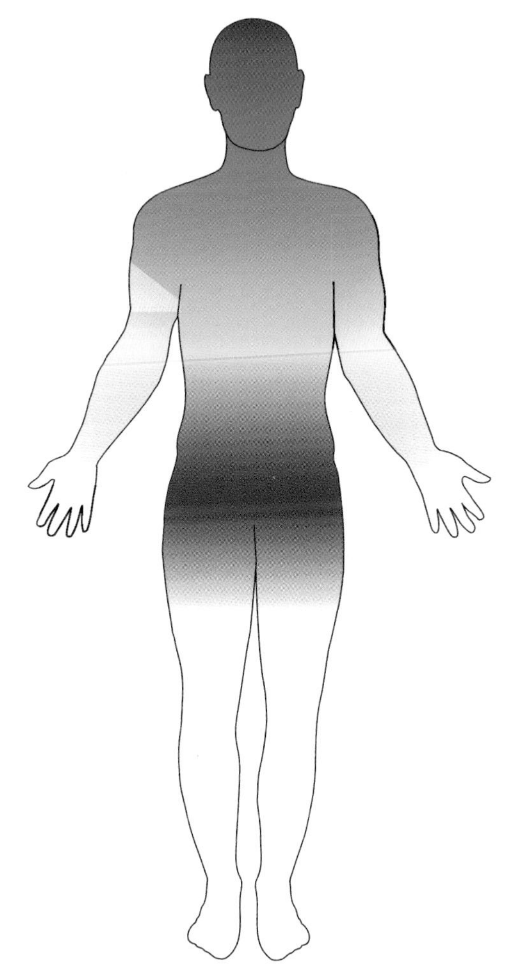

A daily meditation

I am who I am,
down to the centre of the earth
and up to the stars.
And in my heart I am who I am
and none other.
There am I green and magenta.
And I feel linked
to all locations
which emit strength.

Magenta is not really a chakra colour, but if necessary it can be associated with the heart chakra. Magenta is even greater in degree than white light, for it is the intensification and enhancement of all the colours.

You experience wholeness when you include colours in your meditative consciousness or in your dreams. But the danger in doing this is that you might begin to neglect the world of the senses. After undertaking a meditation involving colours the best thing you can do is go and have a good meal that is as colourful as possible. This is yet another way of taking colours into yourself.

Colour for healing with Marko Pogacnik

Marko Pogacnik, the well-known geomancer from Slovenia, has shown in his earth-healing seminars that visualized colours can also be brought to bear effectively. The following text is taken from his book *Healing the Heart of the Earth*.

The power of colours

Regarding the meaning of colour in earth healing, Devos, the Angel of Earth Healing, transmitted a message ... on 14 January 1996 in which he said:

'The entire world is made of colour, and so is the energy plane of space. All energy structures you know have a specific colour vibration. In reality, colour is the basic vibration which lends an energy system its form, function, meaning etc. It is part of the function of colour that each structure has a basic main colour to characterize it. There is no fixed rule that says which colours certain systems display. It depends mainly on the quality of the space in which the system is situated.

'One could say that differences in colour quality make it possible to differentiate between various energy centres or systems of space. The specific colour represents the individual element by which an energy system or phenomenon is integrated into the totality of the space. One could imagine this as each of these centres or systems not always but usually being composed of several colours – that is, a ground colour representing the foundation and accompanying colours that belong to the totality of the system, denoting its pattern or added qualities. It is important to know that it is colours that constitute the vibrational pattern and therefore lend a common background to all energy structures. It often happens that the colour which is used in healing has to do with the ground colour of the energy system that is being healed.

'A second way to implement colour in healing consists of applying certain colours or colour vibrations to stimulate helpful processes such as transformation or cleansing. As different colours have different vibrations and properties they are excellent for healing rooms, people, plants, elemental beings and so on. The principle of this kind of healing rests on the interference, i.e. the mutual resonance, between the colours that characterize the object of our healing and the colours that we use to assist us in the process. Often just a minimal correction of colour nuance is needed in the object of healing and can be achieved in this way. Now to the individual colours:

'– White embodies the quality of perfection and purity and is therefore excellent for raising the energy level of a place, person or whatever else.

'– Purple [magenta] is a combination of two opposite vibrations (yin-blue and yang-red) and therefore has a strongly polarizing structure which makes it suitable for cleansing and protection purposes.

'– Blue has gentle vibrations and is able to calm a space and at the same time strengthen it. It fortifies energy structures, but is mainly important for its calming effects.

'– Green is the colour of heart energy. It is very appropriate for healing because it contains the heart vibrations – that is to say, it is centring and at the same time strengthening. Green also has a strong influence on the emotional level.

'– Yellow, too, has a strong effect on the emotional level; on the energetic level it helps in cleansing because its very "sharp" and strong vibrations cause it to be penetrating and concentrated. This is why it can be used when entering along blocked lines after the cleansing process. It can be used to great effect in re-establishing a relationship between two places that has been lost, and it is also suitable as a tool for guided visualization because, as mentioned, it is so concentrated and direct.

'– The vibrations of the colour orange are largely terrestrial, therefore orange is not used very much when working on the energy level. But it has its own properties which can have an influence on the energy plane. For instance, it is used for grounding a space.

'– Red is even more suitable for this purpose, as it has yet more of a terrestrial quality. Because it feels so earthy, it is too aggressive, too strong, for healing purposes and is only rarely applied.'

In the methods of earth healing that Devos conveyed to us, colour is mostly implemented through visualization. The group stands on a certain point and for a period of time visualizes projecting the chosen colour into the space. Purple [magenta], for example, can be used in different ways, individually as well as in group work:

– For example you imagine that the space to be treated is filled with purple [magenta]. In order to stimulate processes of transformation, the colour is not imagined as static, but as in a continuous whirling motion, skipping about in the space. This type of colour visualization can also be accompanied and supported by cleansing or acupuncture singing.

– To purify a space, the area concerned is filled with purple [magenta]. The person visualizing focuses his or her consciousness on the request that the disturbing vibrations of the space may be coupled with the purple [magenta] colour. Then a second stage of visualization follows where a random spot in the space is chosen and imagined as white. Emerging from the white spot one then spreads a translucent white colour in a star shape in all directions throughout the space – slowly and step by step – until the entire space is filled with white light and every last bit of purple [magenta] has been transformed. This form of colour visualization is often used at extremely difficult places as a kind of preparatory cleansing before the group proceeds with cleansing singing.

– In the third procedure too the space is first thoroughly imbued with purple in order to attach disturbed vibratory patterns to it. Then piece by piece the purple colour is 'turned over' and transformed into white. This is another method to stimulate processes of transformation.

From: Marko Pogacnik, *Healing the Heart of the Earth*, Findhorn Press, UK, translated from German by C. Splettstoesser.

A meditation on experiencing colour

by Rudolf Steiner, 1 January 1915

Rose-violet brings us mercy and compassion.

Red: I am permeated with the substance of divine wrath. This wrath is directed against sinfulness. The Last Judgement. I learn to pray.

Orange seeks to arm us with inner power. It wants to strengthen us, to awaken a yearning to recognize the inner nature of things.

Yellow: I live out of the powers from which I was created in the first earth incarnation.

Green: I incarnate into the Now. I become healthy. I become egoistic. I become a microcosm.

Blue: In blue I conquer egoism. I become a macrocosm. I pour myself into the world.

Violet-rose: I sense compassion and mercy in the violet-rose.

'Colour is the soul of nature and of the whole cosmos, and we participate in this soul when we enter into the experience of colours.'

Rudolf Steiner

THE COLOUR ORGANONS

An organon is an all-encompassing representation of a phenomenon. It is a depiction of inter-connections. When the colours are placed within a context, i.e. when an analogy is being sought, this is not a clear, exact science. It remains a preliminary attempt at discovering the associations.

The seven planetary colours

Colour associations for the seven (classical) planets have existed for a very long time. Rudolf Steiner, for example, spoke of several different associations.

In the method of colour therapy practised by the Lukas Clinic at Arlesheim, Switzerland (an anthroposophical clinic treating cancer patients), a patient is irradiated with a different colour every day. He or she remains fully clothed and sits in a dark chamber looking towards a niche which is flooded with coloured light. After a few minutes the colour is extinguished, and the patient then experiences the complementary after-image colour as a flowing ethereal colour which is intended to activate the person's life-forces. A new colour is applied each day, namely that of the planet relevant to the day of the week.

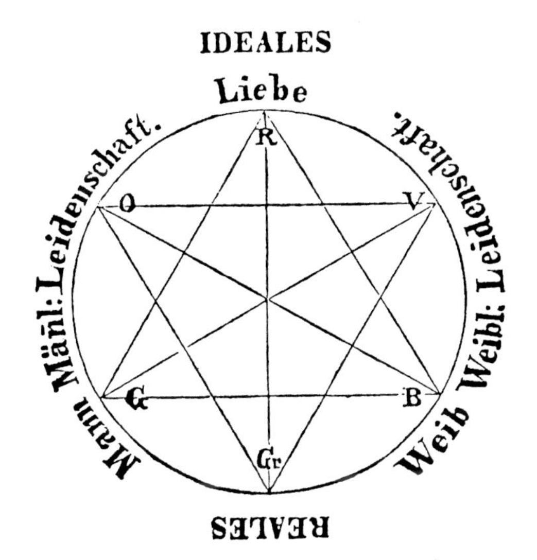

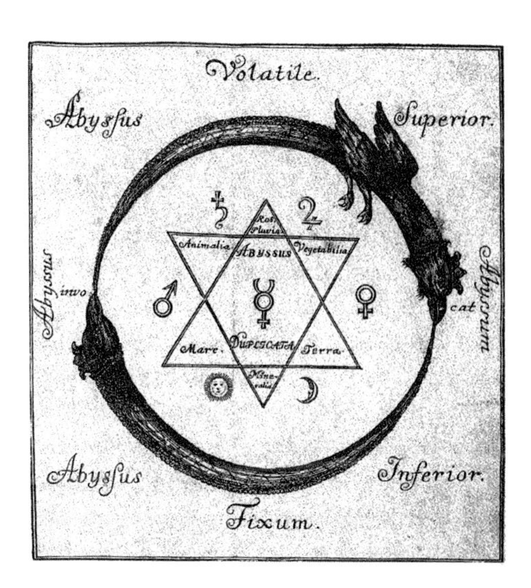

Left: Female and male, the real and the ideal in the colour circle. Right: Colours and planets in alchemy (from: Reinhold Sölch, *Die Evolution der Farbe*, 1998).

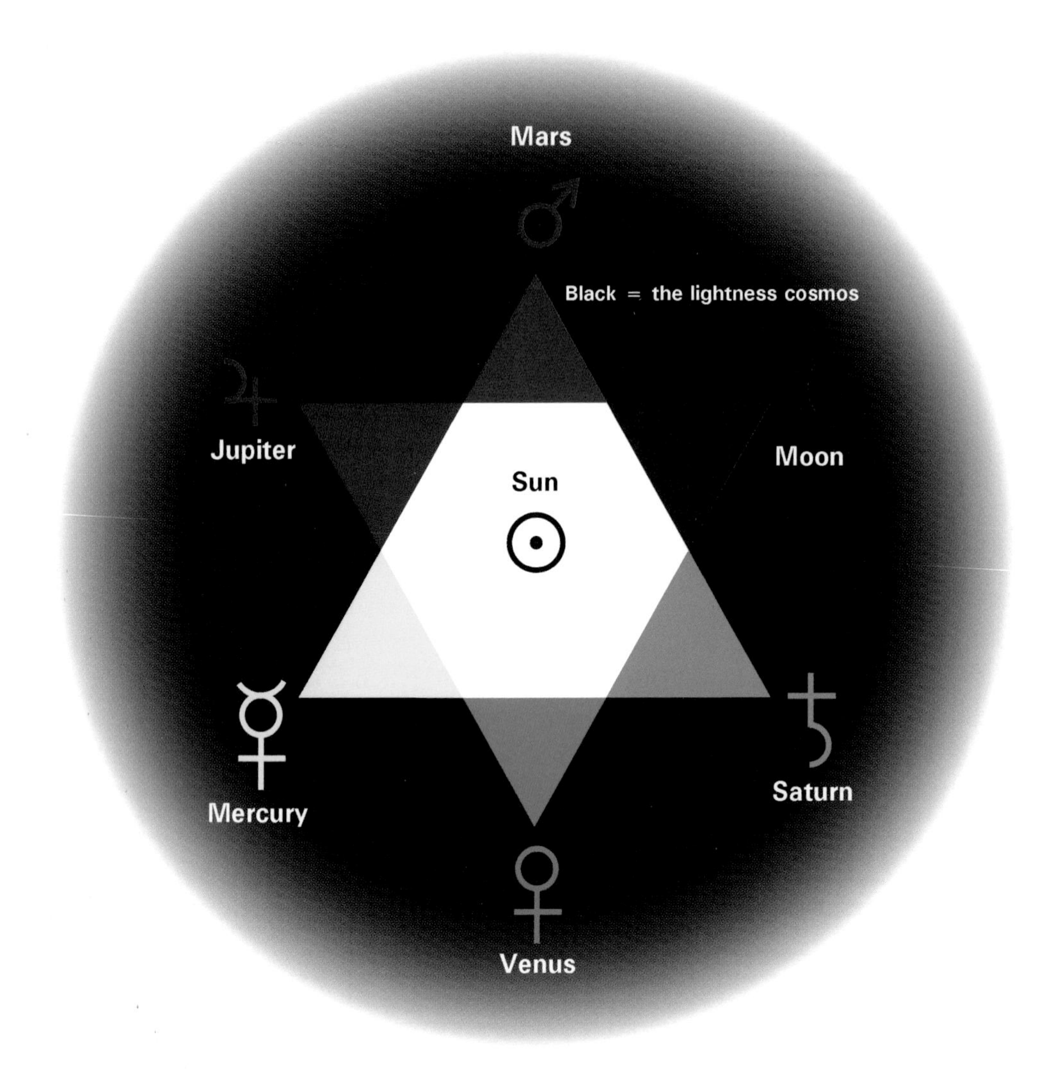

Allocation of the seven planets to the six colours and white. Black forms the cosmos in which the planets move in their orbits.

Poems containing predominantly the sounds of speech linked with that planet are then recited to the patient. Finally, harmonies are played on appropriately tuned lyres. And the patient is given samples of relevant minerals to hold.

The organon designed by me shows more or less the colour analogies as practised at the Lukas Clinic. We shall now endeavour to link the colours of Goethe's colour circle with modern colour theories, for instance that of Harald Küppers. In one of the triangles

Quæ ſunt in ſuperis, hæc inferioribus inſunt:
Quod monſtrat cælum, id terra frequenter habet.
Ignis, Aqua et fluitans duo ſunt contraria: felix,
Talia ſi jungis: ſit tibi ſcire ſatis!
D.M.àC.B.P.L.C.

Alchemical depiction of the stars between light and darkness (from: Reinhold Sölch, *Die Evolution der Farbe*, 1998).

we have the three secondary colours magenta, yellow and cyan blue. In the second there are the three primary colours, vermilion, green and violet, which are at the same time mixtures of the three secondary colours. White and black then join the six basic colours.

We immediately get into difficulties when trying to associate these eight colours with the planets because only since Goethe's time has a distinction been made between two different shades of red. In fact Goethe spoke only of magenta and yellow-red; the latter is here termed vermilion. Even today many people take only one red into account, namely more like the vermilion shown here.

Yellow is generally the most clearly defined, namely as a yellow that contains neither red nor blue.

Blue is more problematical. Not many people would consider the pale shade of cyan blue to be a proper blue, which is normally identified as being deeper and more violet. But cyan blue is the only blue that does not contain any red or yellow. In the same way, magenta is that shade of red which is devoid of any yellow or blue.

Vermilion is the exact mixture of magenta and yellow, as green is that of yellow and cyan blue, and violet that of magenta and cyan blue. Scientifically, violet becomes violet blue.

As seen by the eyes, white and black are always relative. White always has some darkness in it and black some lightness. Shades of grey lie between perfect white and perfect black. In what is perceived by the senses we have practically only shades of grey.

So when endeavouring to associate these colours with the classical planets the most likely correspondences are as follows:

It may be rather surprising to find Mars associated with magenta since Mars is normally seen as more blood-red and vigorous than the spiritual colour of magenta. But magenta is the pure red which works into the realms of blue and yellow and complements the green of Venus in the opposite pole. Communicative Mercury is associated with yellow and introverted Saturn with proper blue. Jupiter which revels in the senses takes its place in vermilion while Moon with its two aspects of light and dark occupies violet. White represents the light-filled Sun, and black stands for the universe, the womb of all creation.

Taking the path followed by the planetary days during the course of one week (i.e. the sequence of colours also followed by patients at the Lukas Clinic), we arrive at this image:

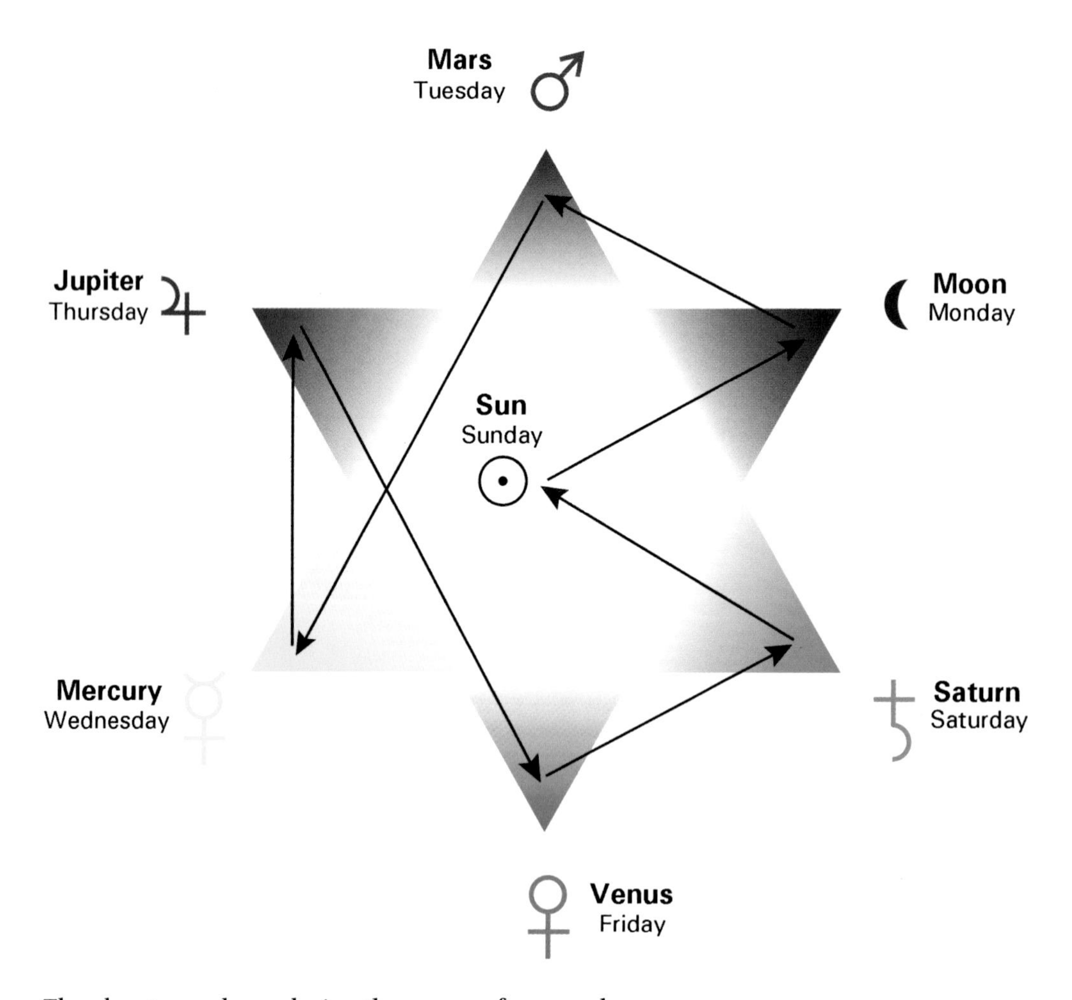

The planetary colours during the course of one week.

Mars sets out from magenta Tuesday and moves through the side of the active colours. On green Friday Venus creates balance and harmony for the passive side of the colour circle although this is interrupted by the central white of the Sun.

When the planetary colours are arranged in a circle, in the Ptolemaic astronomical sequence, interesting interrelationships arise. Neighbouring and complementary colours appear symmetrically. The arrows show the sequence of the week-day planets.

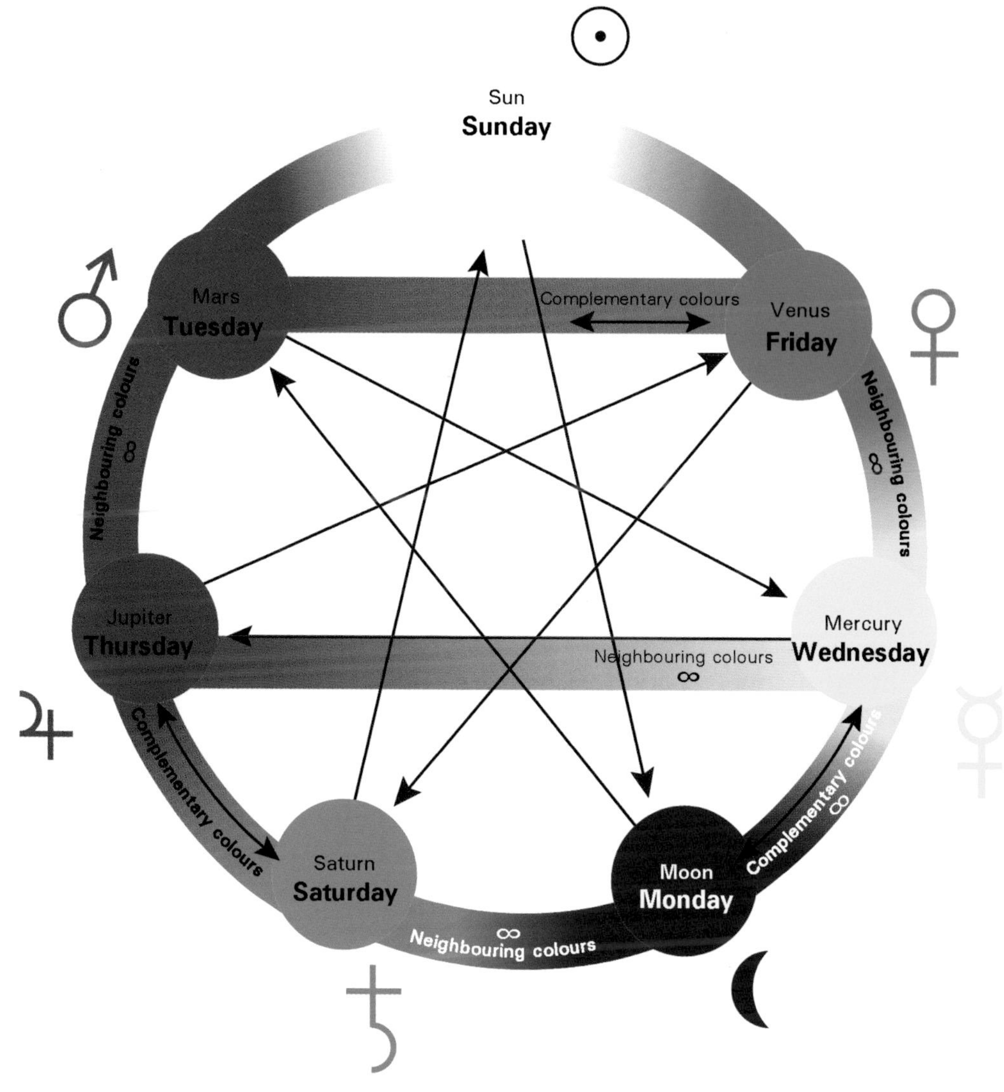

The planetary colours in the heptagon in the Ptolemaic sequence of the week.

The twelve zodiacal colours

It is not easy to associate the colours with the signs of the zodiac because it involves trying to make the zodiac concord with the continuous colour circle. We shall once again make use of Goethe's circle in which magenta is of central importance.

Magenta represents both the beginning and the ending of his colour circle; it rises above the spectrum of the colours between vermilion and violet (see page 62). This makes it obvious that magenta must be placed both at the beginning and at the end of the zodiac. Although Aries is allocated to Mars, we have the all-embracing Sun in the constellation of Aries which then becomes the Sun force itself. Once we have allocated the Aries point to magenta, we still have to clarify the direction in which the zodiac is supposed to move through the circle. It is clear, of course, that the warm colours must be associated with spring, summer and autumn and the cool colours with winter. This yields the following correspondences:

Looking at the zodiacal organon, we see how up to green at the mid-point in Libra the colours grow warmer and more earthly; and then they become more spiritual in the blue, indigo and violet and on to the magenta of the spring equinox, which marks the point from which the coming Easter is calculated. At Easter, the pre-eminent Christian festival, the circle of the year then reaches its culmination in magenta.

The ethereal green of Venus in Libra is the earthly, Michaelic reciprocation to magenta.

The warm orange colour of Cancer on the longest day (St John's) looks across to indigo on the opposite side of the circle, the deepest colour on the shortest day of the year (Christmas). The pure, spiritual yellow of the Sun in Leo looks towards the transcendent violet of Uranus belonging to Aquarius. And mercurial pale green Virgo is the polarity to the violet-magenta of self-sacrificing Pisces.

Taurus standing firmly on the earth has the reddest red next to magenta and looks across to Scorpio in its ambivalent turquoise colour.

Gemini in vermilion, standing in ambivalence between yellow and magenta, complements the clear Sagittarius of Jupiter in its pure cyan blue.

Of course we also come up against contradictions. It is in the last resort not possible to mathematize the colours and force them into a fixed scheme of things. Each association is only partially satisfactory although applicable in practice when one wants to find the totality of the colours in, for example, the zodiac.

The colours – the signs of the Zodiac – the senses

For many years a Zodiac Path has been in process of construction at the

Rosenhof Park of the educational establishment Schlössli Ins in the Bernese lake region of Switzerland. Twelve Zodiac Stations now exist. These are linked with the corresponding colours, the four elements, Rudolf Steiner's twelve points of view, and the human being's twelve senses. A botanic educational path, an animal enclosure, springs feeding small streams and ponds, an alchemical rose path, a lithopuncture stone by Marko Pogacnik connected with the landscape temple in the lake region, a herb garden, a bee house and, in the centre, a large arena, are also now present in the park. There is also an astrolabe (5 metres in diameter) for observation of

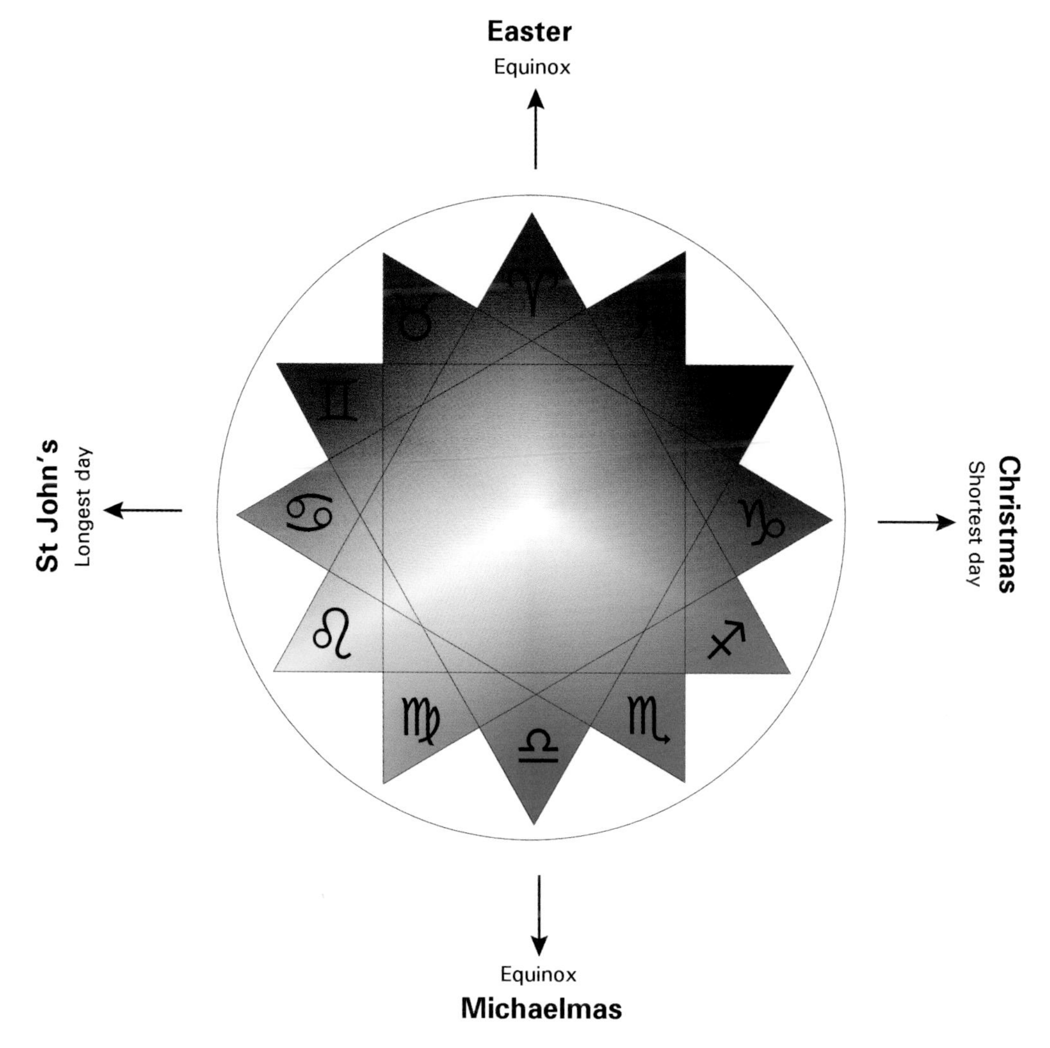

Association of the colours in the zodiac and in the course of the year.

the movements of the celestial constellations. It is painted in accordance with the sequence of colours depicted in this book.

Organon systems also exist which are analogous to the zodiac while being complete in themselves. These invite us to make comparisons. Rudolf Steiner gave various associations for the twelve senses. Here we present the one which has meanwhile become traditional. The zodiac is a continuum with a beginning and an ending. It begins powerfully with the Aries point and ends at the point of Pisces which unites the physical world and the spiritual world, inviting us to pass round the circle once more. The zodiac also represents a path through the human body running from the Aries-like region of the head – which is also indisputably the first to appear at birth – and down to

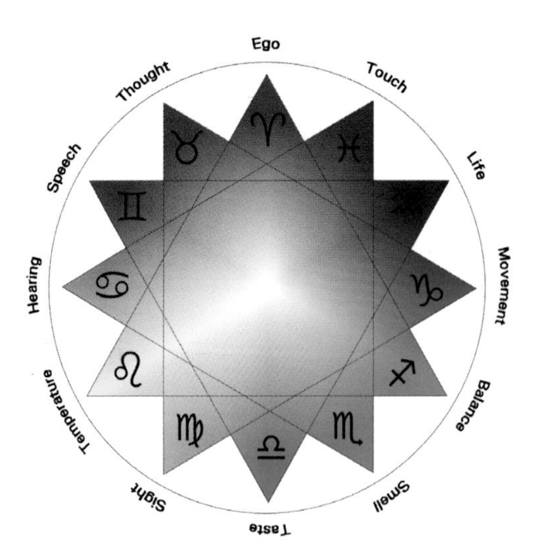

The twelve senses and the signs of the zodiac.

the feet in Pisces. The sequence of the twelve senses begins with the ego sense in a fiery Aries. Only those who feel their own self to be strong can also perceive the uniqueness of the ego in others. In all (sense) perceptions – here the Aries-like forcefulness of the ego – one has to retreat with one's own sense in order to awaken in that of the other. Perception by means of the senses involves entering with one's whole being into that which is perceived. Rudolf Steiner described this as a subtle oscillation between the forces of sympathy and antipathy.

The *sense of thought* is located in the earthy sign of *Taurus*. Since our life of thinking is often conservative in the way it runs along fixed routes, it is symptomatic of our time that perception of thoughts should be connected with the material sign of Taurus. Our sense of thought needs the life forces that can be perceived in the *sense of life* on the other side of the circle. What we find at the Taurus Station in the Rosenhof Park are the compost heaps where material substance is transformed into living forces. As Steiner pointed out, living thoughts lead to Intuition.

The *sense of speech* comes to fruition in *Gemini*. Speech is formed in this mobile centre of communication in the zodiac; not the external mediation of information but speech as it manifests in vowels and consonants, speech as a

The astrolabe at Schlössli Ins which is used to observe the starry constellations.

living being that can be formed through eurythmy. The words Amor, Liebe, Amour, Love, Laska all have the same content and yet each shows a different aspect of love.

The *sense of hearing* lies in *Cancer*. The sign of the introverted spiral winding inwards and outwards shows the sensitive and inspirational understanding of the process of hearing (Steiner). Hearing wants to sense the internal world of experience. We gain a magical and emotional feeling for the world through our hearing.

The *sense of temperature* is associated with the sun-like *Leo* sign. The sun is light, life and love (Steiner). In order to be creative, every process must begin with the warmth of enthusiasm. Warmth is a prerequisite for social

The astrolabe contains an adjustable ecliptic ring which can be set to show the planets tropically or sidereally according to the ephemerides.

processes (Joseph Beuys): In perceiving my own temperature and the temperature of the other (coldness) I also experience the soul element (Steiner). This warmth or coldness of soul is what belongs to the sign of Leo.

As today's most dominant sense, the *sense of sight* is located in *Virgo*. Objectivity, thoroughness and the gift for observation are characteristics of Virgo. Our eyes are gatherers; they graze on whatever they come across in a 'feast for the eyes'. They show us the world as a picture. In contrast to the ears, what the eyes show us is only superficial, the realm of objects as described by the light. Yet perceiving a rainbow, for example, also gives us a foretaste of Imagination (Steiner).

The *sense of taste* is at home in artistic *Libra*. Someone with 'good taste' has a sense of the artistic. In its partnership with Libra, this sense first tastes things externally 'on the tongue' but then immediately takes them into the realm of soul. The Libra Station at the Rosenhof Park is where the pizza oven is situated, an area where sociable feasts for the palate can take place.

The *sense of smell* belongs to the ambivalent sign of *Scorpio*. The sign of 'dying and becoming' unites heavenly fragrances with hellish odours. Rudolf Steiner, born under the Scorpio ascendant (Pisces Sun), links the perception of smell in great detail with mysticism. This is where we become one with God, he points out.

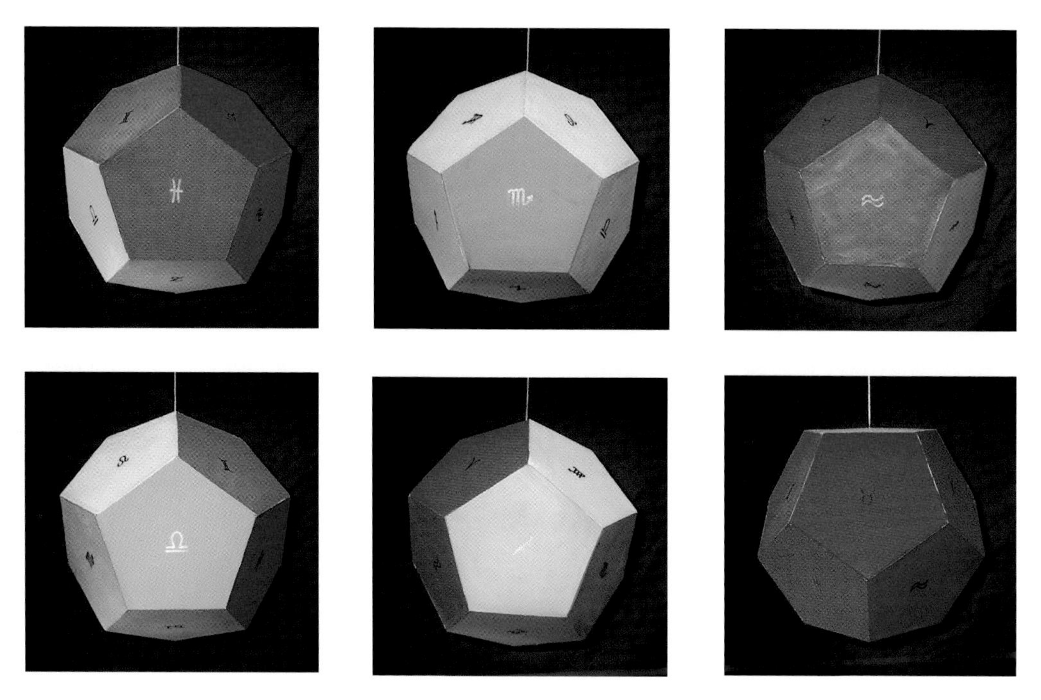

The twelve colours and the twelve signs of the zodiac are arranged so that the colours and signs which complement one another are on opposite faces of a dodecahedron.

The *sense of balance* belongs to ethical *Sagittarius*. This rather fluid moral sense – which is related to subtle hearing – brings about our upright gait. Uprightness, listening to one's own inner voice, allows us to gain a feeling for the spirit (Steiner). It gives us the orientation of an archer with regard to religion, art and science.

It is very surprising to find the *sense of movement* in association with static *Capricorn*, though as an animal the wild goat is, of course, very agile indeed. The knees and the skeleton are assigned to the sign of Capricorn. Capricorn demonstrates the willingness to take on form. Destiny takes place in the ongoing pace of movement. If we let our feet carry us whither they will during a visit to some town or other, we are quite likely to bump into an old friend, perhaps indeed in a fateful encounter. In the sense of movement I experience a soul element that has become free (Steiner).

The *sense of life* belongs to *Aquarius* who carries the water of life. The sign of the inventor, of intuition, of originality, of strokes of genius, calls for an awareness of one's own life energy. The state of 'fluidity' or of the 'white moment' (Daniel Golemann) is a precondition for intuitive creativity. By cultivating our sense of life we are led to spiritual ecology, or to geomancy of the kind practised by Marko Pogacnik.

The *sense of touch* shows us the world of *Pisces*. This sign of mediating, which registers everything intuitively and which puts more trust in feeling than in thinking, is the home of touch. From our sense of touch we learn that the world really does exist. We can only believe that America exists when we set foot on its soil. The sense of touch gives us a feeling of being filled with the mystical reality of God (Steiner).

COLOURS IN MYTHOLOGY

True inner images, like archetypes or primordial images, are deeply rooted in the human psyche. It is no easy matter to describe the inner nature of colours or to characterize them at a psychological or mythological level because by their very nature they have many shades of meaning.

In trying to present some of the ways in which they relate to the world I shall endeavour to follow a more Goethean sequence, one which is more based on processes rather than on a static or random list.

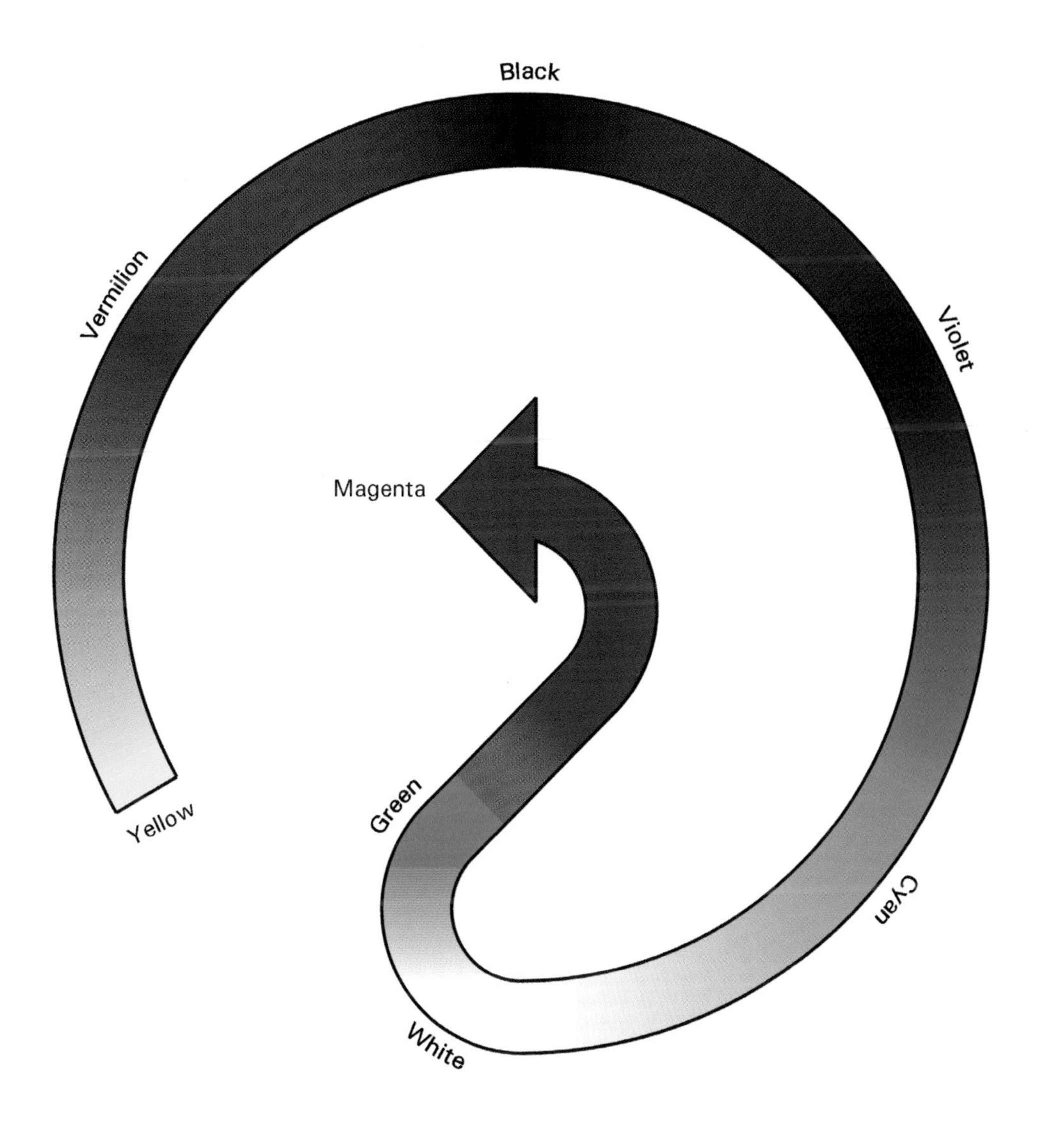

Yellow

When speaking about the colour yellow here we refer solely to pure yellow. Johannes Itten maintains that pure yellow is the only yellow.

As we have seen, yellow is the colour which is closest to light. We associate the rays of the sun and the stars with it. It is the radiance of the spirit, Amon Re in ancient Egypt, Helios and Aphrodite in ancient Greece, Buddha in the (yellow) Middle Kingdom of China, Freya in Germanic mythology, and the Angel Gabriel in early Christianity.

Yellow is both male and female. The ancient Egyptians, but also the Etruscans, painted women in yellow (ochre) and men in red-brown.

Humanity emerged from the yellow-golden age and in the distant future will enter into the joyous new paradise, the yellow-golden 'Heavenly Jerusalem'. Yellow denotes openness, illumination, intuition and the creative force. It is the power of the I, the ego, and of all that is spiritually autonomous.

It has a liberating effect and is extrovert; it can bring the spirit into movement, ranging from a gentle spiritual smile right up to Kandinsky's 'demented laughter'. Yellow is seen as the sublimation of matter, which even for the alchemists meant the transmutation of base metal into yellow gold.

Yellow is the colour of Mercury, of rebirth and communication. Hermes (Mercury) brings human beings from heaven down to earth and after death accompanies them to Hades. Yellow is the creative colour of birth and also the colour of life's harvest in the autumn (van Gogh). Goethe said yellow widens the heart and cheers mind and soul. Is that why harlots, those purveyors of pleasure, were accorded the sensually happy colour of yellow, and is that why they were rejected by the church as being un-Christian?

The liberating yellow of courtly love and spirituality (troubadours) is dangerous for any kind of constraint.

Solid yellow explodes. Yellow wants to move freely, is closest to light. It remains the virginal messenger of the spirit and in yellow light shines in its purest form.

Yellow is the colour of alert intelligence; it is also the chakra colour of the solar plexus as the navel of the world, the 'brain' of the abdomen.

Red (vermilion)

Together with white and black, red is the primordial colour. That 'special juice', human blood, is red; thus red is the existential colour. It is the lustre of what lives (Steiner). It is the first colour of childhood. In Russian the word red also means 'beautiful'. It is the lowest colour of the chakra energy centres and the first in the spectrum after the invisible, warm infrared. Red can be

expansively male and the initial colour in procreation. But it is also the colour of the womb where life begins in warm and living safety.

Red is the primordial fire of Prometheus which gave to humanity the progress of technology as well as independence over against (dark and cold) nature.

Red is the carbuncle sparkling in the dark.

Red is the terror-inspiring colour of the Celtic warrior Ruadh, of Rudra, the Indian god of storms, of Greek Ares. It is the sacrificial colour of killing, of destruction, of chaos, of revolution.

It is also the purifying colour of purgatory which we are obliged to traverse. In Rubedo's chymical wedding the alchemists endeavoured to make the human being blush by combining sulphur and mercury. In Christian art, Mary Magdalene appears in a red garment as one who loves the Lord and the first to see Christ on Easter morning. The red blood of Christ is collected in the chalice of the Grail; and the red wine of the transubstantiation is celebrated in the Christian service of communion.

Red is also the colour of martyrs since only a sanguinary death of the body could qualify them for holiness.

The mythical phoenix burned itself in (red) fire and then rose more beautiful still from the ashes, like the red sunrise ascending from the night's dark depths.

Red is ambivalent for it is not only the Holy Ghost's fiery red flame but also the whore of Babylon and the mother of all earthly abominations.

It is the colour of God's primeval warmth of enthusiasm in the burning bush seen by Moses but also the colour of sex, of the red-light district and of the red-headed witch to be burned at the stake in glowing red fire. It is not only the colour of courageous George but also of the dragon. Red is first and foremost a process colour.

As Goethe said, red arises out of darkened light, preceded in the first instance by spiritual yellow. As the darkness increases we see the sensual erotic orange of courtly love remaining as yet entirely in the realm of soul before descending into the body in the red of cinnabar. In the bodily purposeful emotionality of sex or killing it reaches the culmination of orgasm before fading into the glowing darkness. After 'seeing red' in anger, one descends into despair and crisis.

There is red in the descent into the deep black of death and also in the purification of purgatory, which can even be enhanced into carmine or magenta. This is a progression through the cinnabar-red of hell and on into pure magenta. In a fire we see the opposite process beginning with black coal and progressing via the red glow towards the vermilion flames.

Beginning in the realm of orange we can either proceed to the illumination

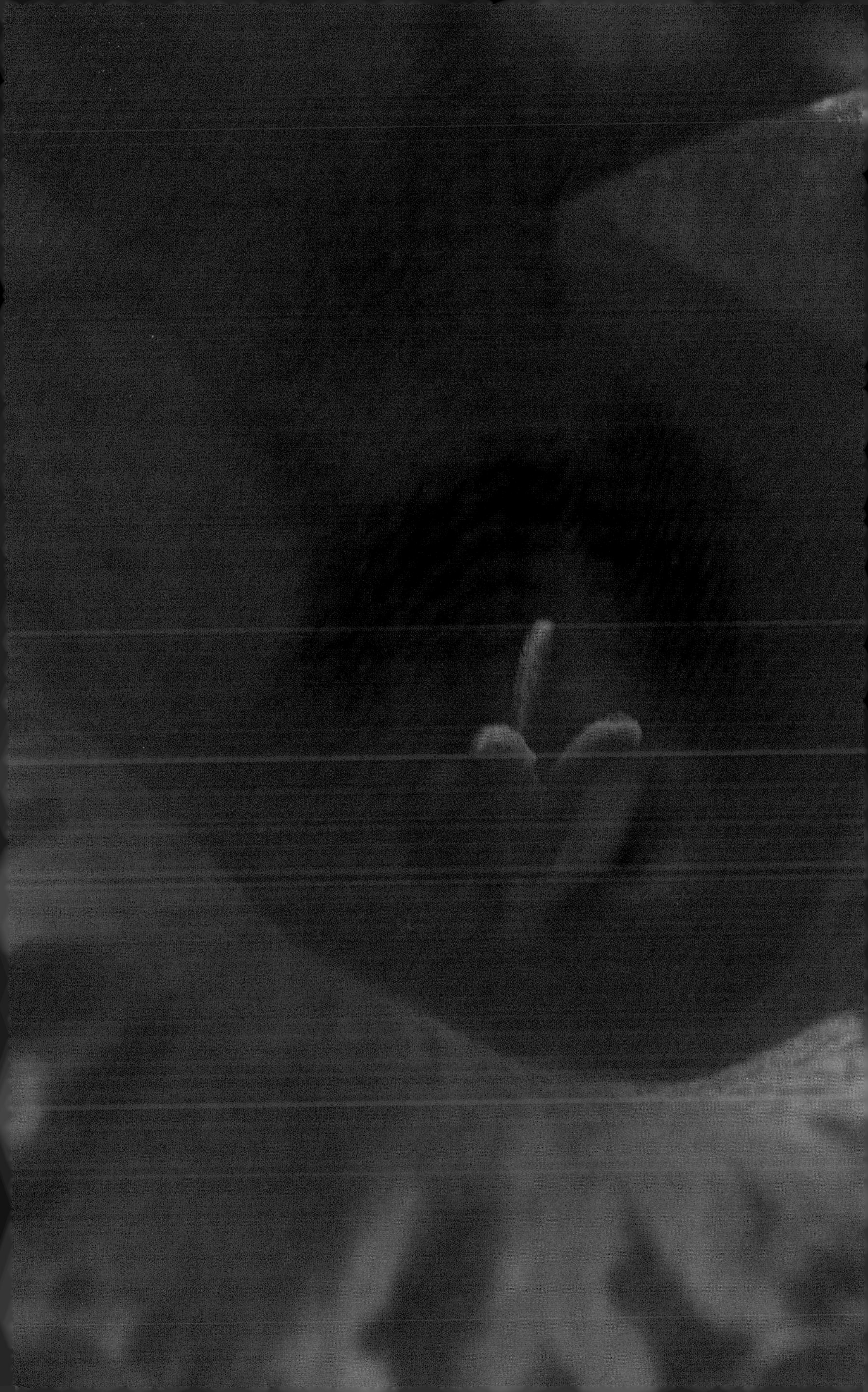

of yellow or burn in cinnabar-red fire. The red witch in the fairy tale casts a spell on Jorinde. Joringel's red flower releases her again. Parzival begins by being the Red Knight before being transformed into the white Grail King. Death in the four-armed cross is elevated to a spiritual essence by red roses.

Black

Black is the sum of all colours in the condensation of subtraction or in negation of all that is light. It absorbs all light into itself as though into a black hole. It is not only an image of our primeval fear, our forsakenness in the dark universe, but also the safe haven of all those who flee. It is a nothingness, something rigid and dead, a corpse (Kandinsky). It is the 'spiritual image of what is dead' (Steiner).

Black is the colour of asceticism, of the intellect, but also of depression and hopelessness.

Black is the 'prima materia', the source of all things. The Black Madonna is the mother of all. To see the sun at midnight entails looking through the blackness of earth in order to experience the spiritual sunlight.

Black can cover over and darken all other colours; but it can also let them shine even more brilliantly when placed beside them.

Black is the magical line which gives painterly motifs their independence and individuality (van Gogh, see page 93).

Black is the borderline colour which comes fully into its own in the nothingness of an abyss. If we see the colour black there must be some light, i.e. white, showing it up.

Black rests our eyes, which even have something of its darkness in their pupils. Through the blackness of our pupils we penetrate into the souls of those with whom we converse. In black nothingness there exists all that is dark, indeed the All itself.

Black is the gateway to hell; it shows us the evil eye but also the spiritual brightness in the glance of those who love.

That which heals and gives life is found in the shadows, in the words of Paul the Apostle. Mary was overshadowed by the Holy Ghost, the primeval source of all that is colourful (coloured shadows). Where there is much light there are many shadows. Light brings form into the shadows. The shadow of our head shows up in the centre of the rainbow, the axis of the dark beam of our looking which – like the eye itself – is surrounded by the brightest aura.

Snow White in the fairy tale is as white as snow, as red as blood and as black as ebony.

Black is the primordial chaos, the primordial slime from which new things can emerge.

In the 1920s Kazimir Malevich's *Black Square* showed that all things must be reduced right down to the 'prima materia' if new processes are to be set in motion. So black is the primordial ground of creation.

Primeval fear can become primeval trust if the darkness of evil, of that which cannot be expressed in words, is integrated into the shadows divided off by the light. Van Gogh needed black as an image of the earth, of heaven, of Saturnine destiny through which not only suffering could be experienced but also, enthusiastically, the coloured world of light.

The stars can only shine forth in the blackness of night. Paul Celan characterized blackness when he said: 'When I am more black in the blackness I am also more naked. When I escape it I become faithful. I am you when I am I.'

Black coal can become a diamond sparkling with light. Light originates from darkness.

Violet

Among the spectral colours violet is exceedingly surprising. Appearing out of indigo, the darkest colour, it lightens itself with red as though wanting to get somewhat closer to the other end of the spectrum where vermilion is situated. Violet is the first colour to emerge from the darkness yet it is lighter than the indigo which follows it. Like vermilion, it is a colour at the edge of darkness. When they are added, these two relatively dark mixed colours yield pure magenta, which contains neither blue nor yellow. So violet cleans the vermilion out of the yellow. Violet is also regarded as a purifying colour in the realm of energy.

Violet is the highest colour in the path of the chakras. As the highest vibration of light, as the crown colour, it integrates the life-filled vermilion with the wisdom-filled blue. In meditative spirituality it cancels out the contrasts. On the crown of the head, in labile balance, it cancels out the profane world of things and transcends it in the meditative colour of mysticism.

It is the dream colour which overcomes space and time in escaping from the material world. As a stranger to reality it becomes embroiled in pondering, losing itself in the elitist idea of 'being better than the others'. Described by Goethe as the 'horror of a universe in destruction' and by Kandinsky as 'sickly', violet can be both stimulating and restraining.

Violet is seen as an androgynous colour in which male and female endeavour to rise to a higher level of humanity. Christ appears in the colour of violet because as a hermaphrodite he has overcome all that is sexual. 'Christ is as though married to himself' (Chevalier).

As an apocalyptic colour, violet characterizes the end of the world just as in the spectral colours it dissolves into the invisible ultraviolet.

Violet is the colour of transformation. It is the colour of rebirth. As the crown colour it leads us into the esoteric world of all that is hidden. And yet paradoxically it is the lowest colour in the rainbow.

Blue (Cyan)

The blue of the 'Blue Flower', from which so much inward warmth and intimacy derived, was the motif of German Romanticism. Blue is a colour which generates profundity. Our eyes can rest in the dark colour of blue since it barely stimulates them to activity. The colour of the 'Blue Rider' is characterized by qualities such as truth, knowledge, wisdom, devotion, faithfulness, melancholia and steadfastness. The colour of sky and sea leads us to an inward peacefulness, indeed it leads us to ourselves.

This colour submerges us in the deep blue of the one with whom we are communicating, with whose mysterious depths we unite inwardly without ever succeeding in entirely fathoming them. The colour of the cosmos teaches us to quieten ourselves in order to listen to the other.

Blue is like the stars – a colour of eternity. This was felt by the craftsmen of the Gothic cathedral at Chartres, who put their feeling for it into those blue stained-glass windows.

Blue as it shines is also a paradox, for according to Goethe it is a 'charming nothing' while also being the colour of the cosmos. Kandinsky, meanwhile, saw it as an unearthly meeting place for all the colours into which one's glance can enter without encountering any obstacle, and thus lose itself in infinity (Chevalier). It is the light of Nirvana, a breaker down of barriers, a quintessence which integrates everything on a higher plane.

On the other hand, blue is the colour of working clothes (overalls), jeans, military uniforms (Prussian blue), and sobriety. Although it shines it is also a colour of darkness. It is the colour of the unconscious but also of alert rationality. It generates fear while also providing security and a sense of safety. It stands not only for visionary fanaticism but also for realistic objectivity. It is the blue of candlelight, of a gas flame and of lightning as it strikes out of darkness. Blue enables darkness to shine. And it is the colour of Europe's star-circled banner beneath which perhaps a united Europe will one day emerge. The scientific definition of true blue, which contains no hint of yellow or red, is cyan. This is a healthy pure blue. In turquoise it moves over towards green and in indigo towards the deepest violet.

Blue has a variety of effects, for example that of lapis lazuli from ancient Egypt, mosaic pieces at Ravenna, the indigo of the oceans, the ultramarine of gentians or the blue of cornflowers. It is the colour of the late Middle Ages. Displacing the golden background in paintings it shone forth as the cloak of the Virgin Mary.

White

White is the furthest point of a lightened darkness. When the deep blue of the sky is lightened until it reaches the whiteness of the clouds, the process of dissolving the darkness has reached its conclusion.

White is both the sum and the absence of all the colours. In the Gospel of Philip, Christ is described as a dyer. Taking 72 pieces of cloth of many different colours, he put them in a cauldron, and when he took them out again they were all white. Mark in his Gospel tells us that Christ's raiment became as white as snow at the transfiguration on the mountain. Mary, immaculate and lily-white, combines within herself pure-white Isis and bright Germanic Freya. Marko Pogacnik saw the goddess of the Seeland region to be a 'White Lady'.

White is the colour of enlightenment, the 'whitening' of the alchemists, the Milky Way in the heavens, pathway of the dead.

According to Kandinsky, white gives us 'the great silence' such as we experience in an untouched landscape deep in snow. White is the light of the illumination attained by the initiates of the 'White Lotus Flower'.

Green

Green reveals itself to us in the chlorophyll of plant life which is brought into being by sunlight. Thus plants become a site where sun and earth, light and darkness, meet. The plant itself therefore becomes an ever-renewing miracle of resurrection, as Simone Weil put it: an image of Christ. And as described by Hildegard of Bingen, this 'nobilissima veriditas', this noblest of green forces, is the Holy Ghost, the power of the heart and the power of Christ. Making ready for the Resurrection on Easter Monday entails preparations during (green) Palm Sunday and during the communion meal on Maundy Thursday (Green Thursday in German). Mary Magdalene was the first to see the Risen Christ as a gardener among the green of nature. So green is the colour of God's Son; it is the colour of reconciliation. And something similar is described in the Old Testament when God created the rainbow after the terrors of the Flood as a sign of the covenant between God and man. It is the green at its centre which makes the rainbow whole by uniting the pale red-yellow with the deep violet-blue.

Osiris, cut into many pieces and made whole again by Isis, was a green god.

Ferdinand Hodler painted his dying friend in green both as she lay dying and as she came back to life. And the Cross on which Christ died is represented as a tree of life in the colour green.

Over and over again, for example among the Celts as well as the Arabs, the spirituality of nature was felt to be green in colour.

Green is the most holy colour of Islam, so that is also the colour of its banner. The mythical 'Tabula Smaragdina' and Goethe's Green Snake are both sources of profoundest mysteries.

In the Middle Ages, green was the colour of courtly love, the colour of the heart. Lying between vermilion, the lowest colour, and violet, the highest, the heart chakra is green.

Green is the colour of the soothing, refreshing meadow; a place, as Goethe put it, where it is pleasant to reside. And according to Rudolf Steiner green is the colour of life.

Magenta

According to Goethe, magenta (what he calls *purpur*) is the pre-eminent colour phenomenon because it unites the two ends of the spectrum. It is a red that contains no shades of yellow or blue, although because it is at the unstable apex of the colour circle it can easily slip towards blue (violet) or yellow (vermilion).

For Goethe, magenta is a synthesis or an enhancement.

Heimendahl describes magenta as 'the healthy centre' between violet and vermilion. It is necessary in our time to make our way from the parts to the whole, just as alchemy was able to proceed from an elementary fourfoldness (here yellow and vermilion together with blue and violet) to a quintessence, namely magenta.

Again according to Goethe, magenta radiates both the gracious dignity of old age and the sweetness and amiability of youth. And for Rudolf Steiner it was a colour in process between light and darkness. As 'peach blossom', magenta is a lightened shade of the blushing flesh tint of the human being.

As poets put it, magenta makes cheeks and lips 'turn red'. Blushing is a soul process of incarnation, a living image of soul forces (R. Steiner). Thus magenta brings to a blushing human being and also to nature the new blossoming of spring. Many buds are magenta and pale green. Magenta and green appear in soap bubbles or in puddles with spilt petrol. A snowy landscape can turn magenta and green at sundown (Goethe).

Earth colour (brown)

The magenta of spirituality is the purest colour. And green is its living complement. But in nature green is rarely pure since it is often interspersed with shades of brown.

Various shades of brown, or mixed colours, are the actual colours of the earth. Plant roots are brown and many-coloured stones give a landscape its identity. The earth colours are mixtures (subtractions) of the pure colours. They delight with their variability. The earth colours complement the pure (rainbow) colours. They contain both the darkness and also the warmth of the earth.

But they are also the culmination of colour densification such as we experience, for example, in the blackness of coal.

In his book *An Outline of Esoteric Science* Rudolf Steiner described the genesis of earlier conditions of the earth. First there was warmth (in what he called the Saturn condition). Then light and darkness evolved in the airy element (the Sun condition). This was followed by a watery development (the Moon condition) in which the colours of the rainbow came into view (Noah's

Ark). The earth colours appeared in a further step of evolution (the Earth 'condition' as such). These earth colours can progress to a high degree of purity which we admire in precious stones, especially tourmaline. In precious stones the dark earth blossoms once more into sparkling splendour. In this sense the earth colours are not only a final culmination but also a new beginning in the process of passing through darkness towards the light.

Bibliography

Aeschlimann, Ueli: *Warum leuchtet die Kerzenflamme?*, Schweizerische Wagenschein-Gesellschaft 1993

Badt, Kurt: *Die Farbenlehre van Goghs*, Dumont Buchverlag, Cologne 1981

Benesch, Friedrich & Bernhard Wöhrmann: *The Tourmaline, A Monograph*, tr. Bruce Allen, Urachhaus, Stuttgart 2003

Bertschi, Thomas: *Der erste Regenbogenkatalog*, Rainbow Project, Schwanden 1999

Bruns, Margarete: *Das Rätsel der Farbe*, Reclam, Stuttgart 1997

Bühler, Walther: *Nordlicht, Blitz und Regenbogen*, Fischer Taschenbuch, Frankfurt a.M. 1982

Burckhardt, Fr.: *Tycho Brahe*, Basel 1901

Burkhard, Ursula: *Farbvorstellungen blinder Menschen*, Birkhäuser, Basel 1981

Collot D'Herbois, Liane: *Light, Darkness and Colour in Painting Therapy*, Goetheanum Press, Dornach 1993

De Witte, Maj: *Ergänzendes zur Farblicht-Therapie*, Verein für Krebsforschung, Arlesheim 1982

Dreyer, J.L.E.: *Tycho Brahe*, Karlsruhe 1894 (reprinted 1999)

Florensky, Pavel: *Iconostasis*, Stuttgart 1988

Gage, John: *Colour in Turner*, Vista 1969

Gebser, Jean: *Gesamtausgabe* (Complete Works), Schaffhausen 1986

Gekeler, Hans: *Taschenbuch der Farbe*, Dumont Buchverlag, Cologne 1991

Glas, Norbert: *Zur Farben-Licht-Behandlung*, Verein für Krebsforschung, Arlesheim 1980

Goethe, Johann Wolfgang von: *Goethe, The Collected Works Vol.12, Scientific Studies*, edited and translated by Douglas Miller, Princeton Paperbacks, Princeton, New Jersey 1995

Goethe, Johann Wolfgang von: *Tafeln zur Farbenlehre*, Insel Verlag, Frankfurt a.M. 1994

Frieling, H.: *Mensch, Farbe, Raum*, Munich 1956

Heimedahl, Eckart: *Licht und Farbe*, Berlin 1961

Heusser, Peter: *Goethes Beitrag zur Erneuerung der Naturwissenschaften*, Paul Haupt Verlag, Berne 2000

Hildegard of Bingen: *Physica*, Vermont 1998

Itten, Johannes: *Kunst der Farbe*, Ravensburg 1962

Jung, C.G.: *Psychology and Alchemy*, Vol.VII Collected Works, Routledge & Kegan Paul, London 1952

Kandinsky, W.: *Concerning the Spiritual in Art*, tr. Michael Sedler, Dover Publishers Inc. 1977

Klee, Paul: *Diaries*, 1957

Küppers, Harald: *Das Grundgesetz der Farbenlehre*, Dumont Buchverlag, Cologne 1978

Küppers, Harald: *Harmonielehre der Farben*, Dumont, Cologne 1999

Lüscher, Max: *Klinischer Lüscher-Test*, Basel 1970

Neumann, Erich; *Tiefenpsychologie und neue Ethik*, Fischer, Frankfurt a.M. 1993

Novalis: *Henry of Ofterdingen*, English translation 1842

Opus Magnum, Prag: Trigon 1997

Ott, G. and H.O. Proskauer: *Das Rätsel des farbigen Schattens*, Zbinden Verlag, Basel 1979

Pestalozzi, Heinrich, *Pestalozzi – The Man and His Work*, by Kate Silber, Routledge & Kegan Paul, London 1965

Pogacnik, Marko: *Healing the Heart of the Earth*, Findhorn Press, Forres 1998

Proskauer, H.O.: *Zum Studium von Goethes Farbenlehre*, Zbinden Verlag, Basel 1985

Ray, Clarissa: *Mit Farben meditieren*, Edition Tramontane, Bad Münstereifel 1993

Riedel, Ingrid: *Farben*, Kreuz Verlag, Stuttgart 1983

Roll, Eugen: *Der Gesandte des Lichts*, J.Ch. Mellinger Verlag, Stuttgart 1976

Schärli, Otto: *Werkstatt des Lebens. Durch die Sinne zum Sinn*, AT Verlag, Aarau 1991

Sölsch, Reinhold: *Die Evolution der Farben*, Ravensburger Verlag, Ravensburg 1998

Silvestreini, Narciso: *Farbsystemem*, Dumont Buchverlag, Cologne 1998

Steiner, Rudolf: *Colour – Three Lectures with extracts from his notebooks*, Rudolf Steiner Press 1971

Steiner, Rudolf: *Colour – Twelve Lectures*, tr. J. Salter and P. Wehrle, Rudolf Steiner Press, Bristol 1992

Steiner, Rudolf: *From Comets to Cocaine*, Rudolf Steiner Press, London 2000

Steiner, Rudolf: *The Philosophy of Spiritual Activity. A Philosophy of Freedom*, Rudolf Steiner Press, Bristol 1992

Steiner, Rudolf: *An Outline of Esoteric Science*, Anthroposophic Press, Hudson 1997

Stone, Irving: *Lust for Life*, London 1990

Stracke, Viktor: *Das Geistgebäude der Rosenkreuzer*, Verlag am Goetheanum, Dornach 1993

Stromer, Klaus: *Farbsysteme*, Dumont Buchverlag, Cologne 1998

Van Uitert, Evert: *Vincent van Gogh, Leben und Werk*, Dumont Buchverlag, Cologne 1976

Verspohl, Theresa: *Entstehung und Geheimnis des Regenbogens*, J.Ch. Mellinger Verlag, Stuttgart 1992

Vogel, Lothar: *Der dreigliedrige Mensch*, Verlag am Goetheanum, Dornach 1979

von Eschenbach, Wolfram: *Parzival*, Penguin Books 1990

Zajonc, Arthur: *Catching the Light. The Entwined History of Light and Mind*, Oxford University Press 1993

Picture Sources

Thomas Bertschi (Rainbow Catalogue) 28, 42 top

Ruedi Büchler 8, 17

Rene Bürgy 25, 32, 33, 34, 36, 37, 38, 39, 44, 45, 46, 47, 48, 51, 52, 70, 72, 73, 74, 76, 77, 79, 81 bottom, 88, 90, 91, 113, 114, 115, 119

Nicolas Kyramarios 6, 7, 13, 14, 15

Ulla Mayer Raichle 42 bottom, 43, 55, 63, 82, 83, 85, 86, 88, 90, 121, 122, 126, 127, 128, 130, 131

Alma Seiler 9, 11

Julian Seiler 10, 19, 20

Ueli Seiler-Hugova 16, 18, 21, 22, 35, 61, 75

Recklinghausen Observatory 5

Cornelia Ziegler 58